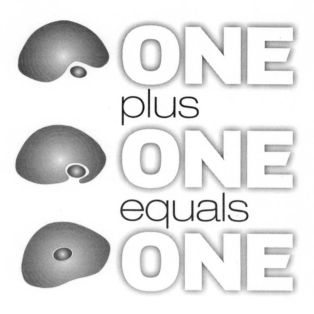

ONE plus ONE equals ONE

symbiosis and the evolution of complex life

JOHN ARCHIBALD

OXFORD
UNIVERSITY PRESS

OXFORD
UNIVERSITY PRESS

Great Clarendon Street, Oxford, OX2 6DP,
United Kingdom

Oxford University Press is a department of the University of Oxford.
It furthers the University's objective of excellence in research, scholarship,
and education by publishing worldwide. Oxford is a registered trade mark of
Oxford University Press in the UK and in certain other countries

First Edition published in 2014

Impression: 1

Published in the United States of America by Oxford University Press
198 Madison Avenue, New York, NY 10016, United States of America

British Library Cataloguing in Publication Data

Data available

Library of Congress Control Number: 2014936760

ISBN 978-0-19-966059-9

Printed in Great Britain by
Clays Ltd, St Ives plc

For Shauna, Cameron, and Miles

Colin Turnbull (1961) took a pygmy friend, Kenge, out of the forest for the first time in his life, and they climbed a mountain together and looked out over the plains. Kenge saw some buffalo 'grazing lazily several miles away, far down below. He turned to me and said. "What insects are those?"... At first I hardly understood, then I realized that in the forest vision is so limited that there is no great need to make an automatic allowance for distance when judging size. Out here in the plains, Kenge was looking for the first time over apparently unending miles of unfamiliar grasslands, with not a tree worth the name to give him any basis for comparison...When I told Kenge that the insects were buffalo, he roared with laughter and told me not to tell such stupid lies.'

Richard Dawkins, 1982, The Extended Phenotype, p. 7

CONTENTS

LIST OF FIGURES

ACKNOWLEDGEMENTS
AND NOTE TO READER

The idea for this book crystallized during the evening of 4 July 2008. I was sitting in a cozy dining room with about 40 other scientists, spouses, students, and friends atop Citadel Hill, a beautiful National Historic Site in the heart of Halifax, Nova Scotia. The occasion was the closing banquet of a meeting entitled Mitochondria, Ribosomes and Cells: A Symposium in Honour of Mike Gray. Professor Emeritus Michael Gray was retiring, and the preceding two days had been filled with talks covering much of the scientific territory he and his lab had roamed over since 1971. As I sat listening to speeches from Gray Lab members past and present, I realized how little I actually knew about the history of my field. I decided to do something about it. The end result is this book, and a much greater appreciation of science as a process. I sincerely thank Michael and my mentor-turned-colleague Ford Doolittle for 15 years of inspiration and support, and for teaching me the importance of looking backwards as well as forwards.

On more than one occasion before and during the writing of this book I was warned of the dangers of telling stories about scientists and the science they do. As with any text dealing with historical aspects of research, not everyone will agree with what I've written. Errors of fact are of course mine alone; so too are errors of interpretation, where possible to discern. I must emphasize the fact that there is much interesting and important science that was carried out during the time span covered in this book that I have glossed over or ignored entirely. Jan Sapp, Professor of Biology and History at York University in Canada, has written extensively on many of the topics discussed herein; I encourage those interested in digging deeper into the history of endosymbiotic theory to consult his numerous articles and books. I have learned a great deal from Jan's work and thank him for his insights.

I have accumulated a great many additional debts over the past two years. I thank Professor Christopher Howe and Adrian Barbrook for being gracious hosts during my sabbatical stint as a Visiting By-Fellow at the University of Cambridge. I am grateful to Chris, Adrian, and all the Howe Lab members in the Department of Biochemistry for discussions and companionship. Churchill College served as home base for much of the front-end research for this book, and helped make my 'Cambridge experience' rewarding and inspirational in ways that I hadn't anticipated.

I am greatly indebted to Oxford University Press (OUP) and various members of their team for support, advice, and encouragement. First and foremost I thank senior commissioning editor Latha Menon for agreeing to take this project on. Latha's broad scientific knowledge and expert editorial advice helped make this book a better, crisper final product. Emma Ma, senior assistant commissioning editor at OUP, is also thanked for her efforts, particularly as the book transitioned from draft through to production. Finally, I thank former senior publisher for journals at OUP, Cathy Kennedy. Early in the project, before most of the book had been written, Cathy provided sage advice on what popular science writing should and should not try to achieve. She also provided valuable comments on early chapters.

Numerous additional people, scientists and non-scientists alike, read some or all of the text in draft format (in several cases more than once). I am immensely grateful for their encouragement and critical comments. In alphabetical order, these people are: Martha Archibald, Shauna Archibald, Andrzej Bodył, Bruce Curtis, Ford Doolittle, Mark Farmer, Martin Embley, Laura Eme, Victor Fet, Gillian Gile, Michael Gray, Michelle Leger, William Martin, John McCutcheon, Michael Melkonian, Nancy Moran, Thomas Richards, Courtney Stairs, and Max Taylor. I also thank Lee Wilcox for help with illustrations and Eunsoo Kim, Eva Nowack, and Takuro Nakayama for providing micrographic images.

I communicated with a great many academics during the research and writing of this book, everything from one-off emails to lengthy face-to-face interviews. Several people deserve special mention. Professor William Martin of the Botany Department at Heinrich Heine University in Düsseldorf, Germany, was incredibly generous with his time, providing detailed insight into endosymbiosis research as well as translations from German. I benefited greatly from time spent discussing organelle research in the pre- and post-molecular era with Bill and Professor Klaus Kowallik; both kindly made their extensive reprint collections available during my visit to Düsseldorf, saving me much time and energy. Nick Lane, senior lecturer at University College London, has been a regular source of inspiration and support, and provided sound advice on the structure of the book in its early stages. Professor Tom Cavalier-Smith and Professor Emeritus Philip John graciously accepted me into their homes for interviews during a visit to Oxford in December 2012. Professor Emeritus Max Taylor did the same on a trip to British Columbia (Max also kindly shared portions of his memoirs with me), as did Francisco Carrapiço and Ricardo Santos on trips to Portugal. Francisco and Ricardo provided me with interesting perspectives on the nature of science and the careers of numerous scientists, in particular Constantin Mereschkowsky and Lynn Margulis. For additional insight, through their publications and/or in person, I thank the following individuals: John Allen, Steven Ball, Debashish Bhattacharya, Neal Blackstone, Andrzej Bodył, Linda Bonen, Sam Bowser, Donald Bryant, Martin Embley, Paul Falkowski, Mark Farmer, Viktor Fet, Arthur Grossman, Kwang Jeon, Andrew Knoll, James Lake, Antonio Lazcano, Brian Leander, Uwe Maier, John McCutcheon, Geoff McFadden, Michael Melkonian, Maureen O'Malley, Eva Nowack, Thomas Richards, Andrew Roger, Mary Beth Saffo, Jan Sapp, Joseph Seckbach, Alastair Simpson, and Mitchell Sogin.

I sincerely thank my family for unwavering support and encouragement during my sabbatical. Finding quality time to think and write proved to be surprisingly difficult. But at the end of the day a bit of Harry Potter with Cameron and Miles never failed to put things in perspective. My wife Shauna, my parents, and my siblings were always enthusiastic about where I was and what I was doing, even when I wasn't. I am grateful to the members of my lab at Dalhousie University for their independence and hard work in my absence. My lab manager, Marlena Dlutek, is thanked for her endless literature searches on my behalf and for 'running the show' when I was physically not around and/or had my head stuck in book land. For financial support during my time abroad I thank The Canadian Institute for Advanced Research, Program in Integrated Microbial Biodiversity, and Dalhousie University.

JOHN ARCHIBALD
November 2013

INTRODUCTION

We are in the midst of a revolution. It is a scientific revolution built upon our understanding of DNA, the hereditary material of life. Using the tools of molecular biology, we probe and prod the world around us in ways unimaginable a few decades ago. Big or small, past or present, no organism is immune. Need to identify and track a bacterium at the root of a hospital outbreak? No problem: the offending germ's complete genetic profile can be obtained in 24 hours. Curious about how we humans differ from our closest relatives, the now-extinct Neanderthals? Anthropologists are tackling this very question with DNA extracted from fossilized bone. With a simple cheek swab and a hundred dollars you can delve into your family history in ways that traditional genealogy cannot, and the same technology can reveal susceptibility to Alzheimer's disease and certain cancers. It can even solve a crime. We insert human DNA into E. coli bacteria to produce our insulin; we isolate spider silk protein from the milk of transgenic goats; we dream of solving the energy crisis with microbes engineered to produce alternative biofuels; we can clone our pets.

Precisely when and how this revolution began is debatable, but it is safe to say that it would not have happened without the advent of DNA 'sequencing'. The iconic DNA double helix is a ladder-like molecule whose rungs are comprised of four chemical 'bases'. In the 1970s scientists did a remarkable thing: they figured out how to take a piece of DNA and determine the precise order of these bases from one end of the molecule to the other. The microbiologist Carl Woese referred to it as 'biology's ultimate technique', and for good reason. Buried within the four-letter chemical read out—within the DNA

1

sequence—are an organism's genes, collectively the set of 'instructions' it uses to synthesize the proteins needed to sustain life at the cellular level. The genetic alphabet is simple but the information it stores is complex and powerful. As we learned the language of life we learned to manipulate it with increasing ease, transforming countless areas of basic and applied research. Science would never be the same again.

It is natural to look at biotechnology in the twenty-first century with an uneasy mix of wonder and fear. Biotechnology is, however, not as 'unnatural' as one might think. Indeed, biotechnology would be utterly impossible if not for the following indisputable fact: all life on Earth is related. All living organisms use the same fundamental molecular processes to maintain and replicate their genetic material; all organisms use the same basic genetic code to 'read' their genes. From wombats to whales, yaks to yeasts, barnacles to bacteria, the similarities are written in their DNA. And here's the astonishing thing: evolution has been 'plugging-and-playing' with the molecular components of life from the very beginning, generating new organisms with novel biochemical capabilities. It continues to do so today. The evidence is there for all to see. In fact we *are* the evidence; we need look no further than the inner workings of our own cells.

Molecular biology has allowed us to gaze back more than three billion years to the ancient roots of unicellular life. It has revealed how, from simple precursors, complex life forms came into existence. This book tells the story of how we have come to realize that our cells are natural chimaeras, and the importance it holds for us as human beings. It is some of the most exciting and important scientific detective work most people have never heard of.

1

LIFE AS WE DON'T KNOW IT

The yin and yang of life

S pring, summer, autumn, and winter—the temperate seasons of Earth. The further we are from the equator the more extreme they become, and the more our lives turn with them. Our blue-green gem of a planet orbits the sun at a breathtaking 30 kilometres per second. It is a journey that takes 365 and one-quarter days. And all the while Earth spins like a top, its axis of rotation tilted 23.5 degrees relative to its orbital plane. It may not seem like much of an angle, but the northern and southern hemispheres alternate summer and winter precisely because of it. In summer, the days are longer and the sun's rays strike the Earth more directly. It's a simple matter of physics. From the perspective of biology it makes all the difference in the world.

Nothing serves to focus the mind on the wonders of nature quite like a walk. Any season will do, although a sunny day in late spring or summer is apt to be most enlightening. If you're so inclined, pretend you are Isaac Newton and sit under a tree. Maybe it's an apple, or perhaps a majestic oak. Relax. Lean back, stare upward, and consider what's happening. If the tree is deciduous, its branches will be covered with leaves, a sea of green against a bright blue sky. Inside those leaves the world's most important biochemical reactions are taking place. With little more than carbon dioxide, water, and some nutrients drawn up by its roots, the tree is harnessing the energy of light, light that has travelled 150 million kilometres from the sun. Individual particles of light—photons—pass cleanly through the waxy outer layers of its leaves and on towards specialized structures called *chloro-plasts* buried within the cells beneath. Inside these sub-cellular factor-ies, which give leaves their colour, photon absorption by chlorophyll pigments triggers a series of reactions that ultimately produce sugar and release oxygen as a by-product. This is the process of *photosyn-thesis*. With minor variations it is happening pretty much everywhere: within the plants on your porch, inside the needles of towering dawn redwoods, in slimy beach kelp, in desert cacti, even in single-celled algae drifting aimlessly in the sea. Photosynthesis makes the world go around.

The collective efforts of these and a myriad of other photosynthetic organisms have, over many millions of years, shaped the chemical composition of Earth's atmosphere. Their influence can still be felt. Carbon dioxide levels wax and wane as day turns to night and back again—as its name suggests, photosynthesis does not occur in the absence of light. During spring and summer, with leaves aplenty and the engines of photosynthesis running full tilt, average atmospheric carbon dioxide levels gradually decline. But with the arrival of autumn, and the transition from green to red, orange, and brown that accom-panies it, rates of photosynthesis slow: carbon dioxide accumulates once more. One could be forgiven for assuming that the northern and southern hemispheres cancel one another out in this regard, alternat-ing summer and winter as they do. Not so. Because there is more land mass and more vegetation in the northern hemisphere than south of the equator, its collective thirst for carbon dioxide is greater. Conse-quently, global carbon dioxide levels fluctuate on an annual basis as a result of photosynthesis. Year in year out, levels decline when the North Pole is tilted towards the sun, and rise again when it tilts away. As we breathe, so too does our planet.

Leaves, pine needles, and kelp blades make sense to us. We can see them, touch them, and easily picture their flat surfaces catching light like solar panels on a roof. Chloroplasts? More than a thousand of them could fit on the head of a pin. What exactly are these invisible light-eating factories, and where did they come from? Despite being integral components of the plant and algal cells in which they reside, chloroplasts were once masters of their own destiny. They used to roam free. We know this because their free-living ancestors are still with us. Well, sort of. A particular group of bacteria, *cyanobacteria*, also make a living by carrying out photosynthesis. They are pretty good at it too. In fact they invented it, somewhere between two and three billion years ago. And having evolved the ability to feed themselves with inorganic compounds and a dash of sunlight, cyanobacteria took another step. It was an incredibly important step, one that ultimately triggered the greening of our planet. More than a billion years ago relatives of today's single-celled cyanobacteria took up residence inside a very different, more sophisticated type of cell, a *eukaryotic* cell. They have been there ever since. The chloroplasts of plants and algae evolved from cyanobacteria by a process called *endosymbiosis*: the intimate association of two distinct life forms, one inside another.

The energy you are using to read this book also came from the sun. Not directly, but as a result of cellular respiration. With each breath, the oxygen liberated by trees, plants, algae, and cyanobacteria makes its way to a different sort of factory within your cells, to the *mitochondrion*. Here oxygen plays an essential role in the conversion of carbohydrates—obtained from your food—into adenosine triphosphate (ATP), the energy currency of living cells. Everyone knows that we humans don't last long without oxygen. A lesser known but equally important fact of life is that our mitochondria are, in essence, domesticated bacteria: they evolved from bacteria by endosymbiosis, just as chloroplasts did. The proverbial apples first began to drop in the late 1800s, but it would be almost a century later before it could be proved that endosymbiosis happened and that it was an evolutionary force that needed to be taken seriously.

Photosynthesis and respiration are intimately related processes, the biochemical yin and yang of life. While the differences between plants and algae on the one hand and food-ingesting organisms on the other are significant, the flow of energy through their chloroplasts and mitochondria connects life in the furthest corners of the biosphere.

In order to fully appreciate the important role chloroplasts and mitochondria have played in the evolution of complex life, we first need to take stock of the diversity of cells in all their minuscular glory. What better place to start than the walking, talking microbial incubator that is *Homo sapiens*. Let's step inside.

It's a jungle in there

Imagine what would happen if everything that is 'you' disappeared. If all the cells of your body were to suddenly vanish, what would be left? This intriguing thought experiment was posed in 1985 by the pioneering ecosystem biologist Clair Folsome. Here is his answer:

> What would remain would be a ghostly image, the skin outlined by a shimmer of bacteria, fungi, round worms, pinworms and various other microbial inhabitants. The gut would appear as a densely packed tube of anaerobic and aerobic bacteria, yeasts, and other microorganisms. Could one look in more detail, viruses of hundreds of kinds would be apparent throughout all tissues. We are far from unique. Any animal or plant would prove to be a similar seething zoo of microbes.*

Although impossible to determine precisely, back-of-the-envelope calculations come up with ten trillion as the minimum number of cells comprising a human being. Ten trillion cells: that's a one with 13 zeros behind it, far too large a number to easily get one's head around. One hundred trillion is even worse, which is the number of *bacterial* cells thought to be living within and on the human body. We are outnumbered ten to one by bacteria alone. The gastrointestinal bug *E. coli* springs quickly to mind, but there are in fact more than 500 different bacterial species living in our digestive systems, another 500 in our mouths, and dozens in our bellybuttons. And it is not just simple, single-celled creatures that we are lugging around. We've got

---

* Viruses and bacteria, fine, but pinworms and fungi? I first heard of Folsome's experiment in an undergraduate microbial diversity course and distinctly recall finding it disturbing. I confess I still do.

eight-legged mites hunkered down in the follicles of our eyelashes, munching away on dead skin. The human body is an ecosystem in which every nook and cranny is inhabited by something other than 'us'. In this sense we are a microcosm of the world we inhabit, a world in which no niche is too small or too extreme to be left unfilled, a world in which organisms readily forge intimate associations with one another if and when the opportunity arises. We could probably get by without the mites, but we are utterly dependent on our microbes, and they depend on us.*

In feeding humanity's obsession with cataloguing the flora and fauna of Earth, the microscope has done for the natural sciences what the telescope did for the physical: it unveiled a world that hitherto did not exist in any meaningful sense. To early physicians trying to understand and improve human health, the microcosm was as vast and mysterious as anything to be pondered in the heavens; it was in the spirit of exploration, not a quest for a cure, that it first came into focus. The 'father of microbiology' was in fact not a scientist at all: Antonie van Leeuwenhoek (1632–1723) was a Dutch draper with a passion for lenses. He certainly wasn't squeamish. In his spare time he built the most powerful magnifiers of the day and used them to examine bodily fluids. Van Leeuwenhoek was probably the first to observe red blood cells and spermatozoa, among other human cell types, as well as a wide variety of microorganisms, or 'little animals' ('animalcules') as he called them, asserting that 'there are more animals living in the scum on the teeth in a man's mouth, than there are men in a whole kingdom'.[1] Together with the Englishman Robert Hooke (1635–1703), coiner of the term 'cell' and author of the best-selling *Micrographia*, van Leeuwenhoek laid important groundwork for the

* Lest we get too carried away, it is important to realize that although our cells are seriously outnumbered, they are much larger and heavier than bacterial cells. Consequently, microbes make up at most only a few per cent of our total body mass; pound-for-pound, we are most definitely human.

eventual development of cell theory: the cell is the fundamental unit of the organism and all cells come from other, pre-existing cells.*

Three hundred years in the making, modern microscopy has become a thing of beauty. Even basic light microscopes can see all but the very smallest of cells, and with electron microscopy the wavelength of visible light (~0.5 micrometres, or half a millionth of a metre) ceases to be a limitation; nanometre- and even picometre-sized objects can be discerned. Microscopists investigate sub-cellular architecture in much the same way anatomists study the tissues and organs of animals, and they ask the same basic questions. What are the building blocks and what are their functions? How much variation is there from species to species? How do the individual parts come together to form an organism capable of growth and reproduction?

The answers not only speak to the common ancestry of all cells, they reveal the deepest, most basic divisions of life. Few things in biology are as clear: there are two—and only two—types of cells on Earth, *prokaryotes* [pro-care-ee-oats] and *eukaryotes* [you-care-ee-oats]. Ungainly terms, these, but not to worry, they'll roll off the tongue soon enough. What matters is their etymology. In Greek, eukaryote means 'true kernel' or 'true nut', prokaryote means 'before kernel/nut'. Eukaryotic cells have a nucleus; prokaryotic cells do not. It's that simple. The nucleus of a eukaryotic cell is a prominent spherical compartment where most (but not all) of the hereditary material—the all-important DNA—resides. For this reason, the nucleus is sometimes referred to as the 'command centre' of the cell. Prokaryotes have

* Van Leeuwenhoek is said to have possessed a copy of, and been inspired by, Hooke's famous book, the full title of which says it all: '*Micrographia: or some physiological descriptions of minute bodies made by magnifying glasses with observations and inquiries thereupon*'. Unlike Hooke, who was an esteemed member of London's Royal Society, van Leeuwenhoek was at first dismissed by 'real' scientists at the time as something of a crack hobbyist. It didn't help that he spoke only Dutch and refused to divulge his lens-making secrets. Nevertheless, van Leeuwenhoek's discoveries were recognized as being too important to ignore; he was elected Fellow of the Royal Society of London in 1680.

DNA too, but it is not physically separated from the rest of the cell contents (Fig. 1).

You and I are classified as eukaryotes because we are made up of eukaryotic cells, along with pretty much every living thing that can be

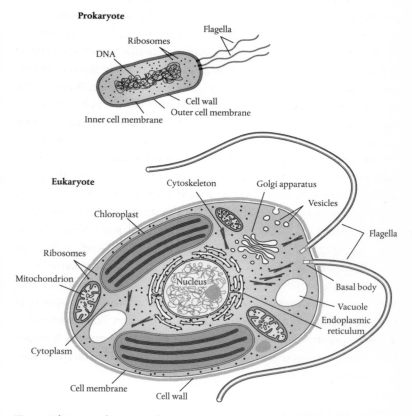

Fig. 1. Schematic diagrams showing cross-sections of prokaryotic (top) and eukaryotic (bottom) cells. Internal structures are labelled. The eukaryote shown is a generic photosynthetic eukaryote; it possesses both mitochondria and chloroplasts, the latter being the light-harvesting organelles of plants and algae. Prokaryotic cells are typically much smaller than eukaryotes (images are not to scale) and possess little in the way of internal complexity. Eukaryotic cells are characterized by the presence of a nucleus, mitochondria, cytoskeleton, and endomembrane system (Golgi apparatus and endoplasmic reticulum).

seen with the naked eye. Every type of animal you can think of (humans included of course), plants and algae, all manner of fungi from baker's yeast to molds, mildews, and mushrooms—all are eukaryotes. Prokaryotic cells are everything else: 'typical' bacteria like *E. coli* and *Salmonella* plus a second, lesser known but important group called 'archaebacteria' or archaea.* Thinking 'big' or 'small' is helpful when trying to decide if an organism is a eukaryote or a prokaryote, but it isn't the be all and end all. Sophisticated multicellular organisms with discrete tissues and nervous systems are invariably eukaryotes, and individual eukaryotic cells are generally much larger than pro-karyotic ones. A human skin cell, for instance, measures about 30 micrometres across, whereas an individual *E. coli* bacterium is only two micrometres long. However, there are 'giant' prokaryotes that can be seen without the aid of a microscope, and a large fraction of single-celled (and microscopic) organisms in nature are in fact eukaryotic. It is what's inside that counts. Single-celled eukaryotes, 'simple' things like yeasts and amoebae, are internally much more like the billions of individual cells that make up an elephant or an oak tree than they are like a prokaryote. There is more to the identification of a cell than what first meets the eye.

Early microscopists, of course, knew nothing of prokaryotes and eukaryotes. Examining thin slices of cork through his 1650s-style microscope, Hooke observed 'microscopical pores' delineated by an irregular network of 'walls'. The pores reminded him of monastery cells, and the walls he saw separating them turned out to be an integral component of plant cells as we know them today. Cells are enclosed by a 'membrane', which is a fat- and protein-rich envelope, thin and flexible, much like the 'walls' of a tent. Plant cells of the sort first viewed by Hooke fortify their surface membranes with proper walls

---

* Prokaryotes are getting short shrift here, and unfairly so. The late Carl Woese (1928–2012), discoverer of the archaea in the 1970s, referred to them as the third domain of life. He considered archaea to be as different from bacteria as bacteria are from eukaryotes. Not everyone agrees. We shall give the archaea the attention they deserve in Chapter 4.

comprised of cellulose and related compounds in order to resist desiccation and changes in osmotic pressure. Cell membranes are incredibly important, and not just because they keep the outside out and the inside in. They are the cell's face to the world, the surface through which they interact with their environment. For multicellular organisms they are the glue that holds them together: cells glom on to other cells to form tissues, tissues make organs, organs work together to make individuals.

In terms of outward appearance, prokaryotes haven't got a lot going for them. Their membranes and (when present) walls are similar to those of eukaryotes, but beyond this there is very little to compare, let alone contrast. Prokaryotic cells are typically oval- or rod-shaped, for the simple reason that they haven't got the infrastructure to support anything more complicated. The eukaryotic cell is a jungle by comparison (Fig. 1). In addition to harbouring the nucleus, the eukaryotic 'cytoplasm' is supported by a dense 'cytoskeleton', a rigid protein-rich scaffold that allows eukaryotic cells to be orders of magnitude larger than prokaryotes and more adventurous in terms of shape. Consider for instance our doughnut-like red blood cells, flattened skin cells, and neurons with their long thread-like projections, three very different types of cells but all found in the same organism. Eukaryotic cells also have an elaborate intracellular cargo transport system, or 'endomembrane' system, which they use to secrete waste, ingest food and drink, and generally move proteins and other macromolecules from point A to point B. Like a city subway system, the bigger the cell the more important the endomembrane system becomes. Prokaryotes have nothing of the sort. And the eukaryotic cytoplasm harbours membrane-bound mitochondria—the energy converters to which our attention shall soon squarely turn—and in the case of plants and algae, chloroplasts, the all-important receptors of solar energy. Cell biologists refer to mitochondria, chloroplasts, nuclei, and various other sub-cellular compartments as *organelles*, 'organs' of the cell.

So the presence of a cytoskeleton and organelles such as a nucleus is what distinguishes a eukaryote from a prokaryote. But there's a lot more to it than that. Eukaryotic cells are capable of 'shape-shifting' in ways that prokaryotic cells cannot, and their innards are highly compartmentalized. Simply put, eukaryotic cells are more *complex*. When I refer to complex life in this book, it is to eukaryotes that I am generally referring. Next to the origin of life itself, the evolution of the eukaryotic cell from simpler prokaryotes was probably the most important event in the history of life. It is also one of the most enduring mysteries in all of biology, a problem made particularly thorny because of the vast and ancient nature of the prokaryote–eukaryote division; scientists know of absolutely nothing in between. Some scientists think that the origin of the eukaryotic cell went hand in hand with the origin of mitochondria and their unique brand of energy conversion. Others suspect that the nucleus, cytoskeleton, and other eukaryotic complexities evolved in a stepwise fashion before the mitochondrion. But as we shall see, thanks to advances in DNA-based research one thing is for certain: endosymbiosis played a major role in the evolution of cellular complexity, first with the eukaryotic cell and then with plants and algae. If eukaryotes had not evolved there would be no life beyond the form of the lowly bacterium. There would be no plants and no animals of any kind. And we wouldn't be here to ponder our existence.

2

REVOLUTIONS IN BIOLOGY

What is life? It's a question for the ages. It is also the title of a book published in 1944 by the Austrian Erwin Schrödinger (1887–1961).[1] It was a short book but an influential one, more than a bit strange, too, coming from a physicist. Philosophical by nature, Schrödinger was a man of many talents. He published on everything from colour perception to cosmology, and received a Nobel Prize for his work on quantum mechanics.* Schrödinger was struck by the orderliness of life, the ability of organisms to grow and reproduce so predictably despite in essence being giant agglomerations of atoms, which physical laws state to be inherently unpredictable. In *What is Life?* Schrödinger framed the question as follows: 'How can the events in space and time which take place within the spatial boundary of a living organism be accounted for by physics and chemistry?' The list of scientists influenced by *What is Life?* is long and impressive, Nobelists of DNA fame Francis Crick and James Watson among them. Not bad for a book containing little in the way of original content—and most of what was original has turned out to be wrong.

In order to fully appreciate what all the excitement was about, let's take stock of the biological sciences in the mid-1940s. Genetics was a mature discipline thanks to over a hundred years of study on heredity

---

* You might have heard of 'Schrödinger's cat', his famous paradox-thought experiment in which a cat is placed in a box with a flask of poison and some radioactive material. Within the framework of traditional quantum mechanics, prior to opening the box and observing whether or not radioisotope decay has triggered the release of the poison, the cat must be considered simultaneously dead *and* alive. Frankly, as a biologist, it's beyond me.

in peas, maize, and fruit flies, organisms in which physical traits could easily be observed and tracked from one generation to the next. Microbiology was still largely 'practical' in nature, and focused on the study of bacteria and their role in human disease. Darwin's principles of evolution—descent with modification and natural selection—were firmly ensconced in the lexicon of biology, driving both theory and practice in diverse research areas (although not yet for the study of microbes). The problem was that nobody knew precisely how an organism's traits were passed on. The 'gene' did not yet have a physical reality. Fuelled by advances in microscopy, the burgeoning field of cytogenetics was increasingly focused on the study of chromosomes, protein-rich structures visible within the nucleus of eukaryotic cells. Evidence for a specific relationship between chromosome dynamics and patterns of inheritance was mounting, but what exactly were these chromosomes and how did they store their hereditary information?

Schrödinger introduced the idea of genes and chromosomes as 'aperiodic crystals', solid structures comprised of irregularly repeating units, and explored different ways in which a 'hereditary code' could conceivably be embedded within them. In doing so he relied heavily upon an obscure paper published almost a decade earlier by physicist-biologist (and future Nobelist) Max Delbrück (1906–1981) and colleagues.[2] Schrödinger was apparently unaware of much of the cutting-edge research taking place at the time, including that of Delbrück himself—a self-proclaimed 'naïve physicist' exploring ideas about organisms, Schrödinger's knowledge of biochemical genetics was woefully inadequate. According to Gunther Stent (1924–2008) of the University of California at Berkeley, Schrödinger's book probably had little to no influence on professional biologists: 'In so far as they bothered at all to read *What is Life?*", they probably considered the title a piece of colossal nerve.'[3]

If biologists paid no attention then who did? Schrödinger's hope, and presumably the inspiration for his one-off foray into biology, was that a detailed understanding of heredity at the atomic level would reveal fundamentally new laws of physics. This did not happen; the laws of chemistry proved to be all that were needed to explain the complex behaviour of living systems. Nevertheless, well into the 1950s Schrödinger's book served as a source of inspiration for post-war physicists eager for more wholesome challenges, an illustration of how the tools of their 'exact' science could be fruitfully applied to important problems in biology. It was during this period that, in the words of Neville Symonds, 'biology stopped being a "sissy" subject and came of age'.[4] *What is Life?* has come to symbolize the birth of a new discipline that emerged organically at the intersection of bio-chemistry, genetics, and microbiology. The initial goal was as single-minded as it was lofty: uncover the secrets of the 'master molecule' of life.

The evolution of molecular biology

Having completed our whirlwind tour of the diversity of life in the opening chapter you might have felt as though something was miss-ing: what about viruses? Fair enough. To say they are abundant is a colossal understatement—we inhale many thousands of viral particles with each breath; a single millilitre of seawater contains millions of them. Curtis Suttle of the University of British Columbia reckons that all the viruses in the ocean stacked end-on-end would extend beyond the nearest 60 galaxies. Yet despite their immense biological signifi-cance, viruses typically don't factor in discussions of cellular diversity because, strictly speaking, they are not alive. Rockefeller University Nobel laureate Christian de Duve (1917–2013) considered viruses to be 'no more alive than is a disk able to play music'.[5] Many viruses lack their own bounding membrane, and when one is present it is not their own; they 'steal' it from their host. While they are comprised of the same building blocks as prokaryotic and eukaryotic cells (proteins, fats, carbohydrates, nucleotides, and so on), viruses are incapable of self-replication. Existing as they do at the hazy boundary between life and non-life, the extreme simplicity of viruses has proven very useful.

Indeed, they have served as the substrate for some of the most important scientific advances of the last century, none more so than the 'Hershey–Chase' experiments of 1952.

The American Alfred Day Hershey (1908–1997) was an esteemed member of the so-called 'phage group',[6] an eclectic, loose-knit group of scientists led by Max Delbrück and united by their interest in bacteriophages—viruses that infect bacteria. Phages were discovered in 1915 by the Englishman Frederick Twort (1877–1950), and touted by the French-Canadian microbiologist Félix d'Herelle (1873–1949) as promising antibacterial agents in the control of disease. But to most researchers at the time phages were little more than an annoyance; left unchecked they could easily wipe out a bacteriologist's hard won collection of cell cultures. Hershey and his talented collaborator Martha Chase (1927–2003) were trying to make sense of a particular virus that attacked *E. coli*. Their genius and creativity was to use DNA- and protein-specific radioactive markers to label the various components of the virus, and a simple kitchen blender to disrupt the interactions between phage and bacterium during infection. Hershey and Chase showed that it was phage DNA—not phage protein—that appeared to carry the instructions for the production of new phage.[7] We now know that a hallmark of viruses is that they make a 'living' by injecting their genetic material into a real cell and hijacking its protein synthesis machinery to make new virus particles. Viruses are the ultimate parasite, infecting all manner of prokaryotic and eukaryotic cells.

The results of the Hershey–Chase experiments are often portrayed as the definitive proof that genes are made of DNA. But a robust connection between DNA and heredity had in fact been made almost a decade earlier. In 1944 the Canadian-American trio of Oswald Avery (1877–1955), Colin MacLeod (1909–1972), and Maclyn McCarty (1911–2005) showed that harmless strains of *Pneumococcus* bacteria could be 'transformed' into ones capable of killing mice by the addition of purified DNA.[8] The result was hard-fought and highly controversial: sceptics wondered whether the purified DNA used in the experiments contained minute traces of protein, and it was the contaminating protein that was the magic ingredient required to pass the virulence phenotype from one bacterium to another.

If aware of the DNA-or-protein controversy at all, most scientists were content to sit on the fence, unconvinced that the concept of the gene, as developed by geneticists studying plants and animals, could even be applied to—let alone informed by—simple bacteria and their pesky phages. Those with a stake in sorting out the true nature of heredity were well aware of the results of Avery and colleagues; they simply had difficulty believing them. Hershey and Chase had been extremely cautious in interpreting their phage-based results of 1952, in part by nature but also because the prevailing view among phage group members was that the complex features of proteins made them a much more appealing candidate for being the hereditary substance. Delbrück for one considered DNA to be a 'stupid molecule', too simple in chemical composition to serve as the master molecule of life.

In the end the Avery–MacLeod–McCarty result stood the test of time, bolstered by Hershey and Chase and a growing mound of data gleaned from different experimental systems. The answer was clear: genes are made of DNA and all organisms have them, no matter how simple or complex. But knowing this did not address the issue of how DNA actually works. Molecular biology can be defined as the investigation of the structure and function of the macromolecules essential for life. It was elucidation of the structure of DNA—right down to the relative positions of each and every atom—that paved the way for a comprehensive understanding of how it functions as the molecule of heredity.

'It has not escaped our notice'

Biologist or not, everyone knows about DNA: it's how we 'pass on our genes'. And because of what it has come to symbolize in modern society—genetics, medicine, biotechnology—everyone knows what it looks like. But DNA is a mysterious and often misunderstood substance because of the complex role it plays in an organism's development and reproduction. DNA—deoxyribonucleic acid—is a long thread-like molecule, rich in phosphorus and nitrogen. 'Thread-like' barely does it justice: it is a mere two nanometres wide, and each of the ~10 trillion cells that make up a human being has about a *metre* of DNA crammed within its nucleus, itself only a few micrometres in diameter (Fig. 2).* All told, our DNA (ignoring that of our resident microbes) written out in letters the size of this text would stretch across the Atlantic Ocean from New York to London.

The human genome—the sum total of the DNA in each of our cells—has ~25,000 genes, which direct the synthesis of more than 100,000 proteins.† Not all of these proteins are present in all of our cells—far from it. By modulating the process of what biologists call 'gene expression', different cells can use the same genome to generate different subsets of proteins, and thus give rise to multicellular organisms with tissues and organs. Yet beneath all this complexity is simplicity. Inherent in the structure of DNA is the answer to the question of how it replicates such that exact copies are passed on to daughter cells upon division, and its simple four-letter chemical alphabet harbours the 'genetic code', which serves as the link between DNA and the proteins that build and maintain cells, the link between genotype and phenotype.

The now iconic structure of the DNA double helix made its debut in the pages of the prestigious British journal *Nature* 60 years ago in 1953,[9] and no doubt 60 years from now James Watson and Francis Crick (1916–2004) will be remembered for many things. Modesty will not be one of them. Their paths to the Cavendish Laboratory at the

---

* There are minor exceptions. Some cells, like mature red blood cells in mammals, lose their nuclei and the DNA within them.
† Scientists still aren't sure how many genes we have. Initial estimates were on the order of 100,000 but were proven wrong by the sequencing of the human genome. We now know that a single gene can give rise to different forms of a given protein, which partly explains how we humans can be so complicated despite having roughly the same number of genes as a fruit fly.

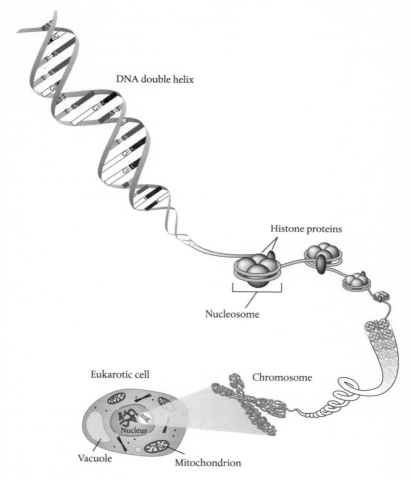

Fig. 2. Schematic diagram showing the packaging of DNA within the nucleus of a eukaryotic cell. The four building blocks of DNA are adenine (A), cytosine (C), guanine (G), and thymine (T). Two DNA strands come together to form a double helix. The double helix is wound around histone proteins to form nucleosomes, which are themselves wound round one another. A chromosome is an individual, highly condensed DNA molecule packaged together with histones and other proteins.

University of Cambridge, the sacred birthplace of molecular biology, could hardly have been more different: Watson, a 23-year-old American phage group disciple trained under Salvador Luria in the late 1940s; Crick, an Englishman in his mid-thirties who had turned to biology after working on 'dull problems' in physics before and during World War II. The building blocks of DNA were by then well known: adenine (A), cytosine (C), guanine (G), and thymine (T). So too was the relative abundance of these four 'bases' in a given sample of DNA, thanks to the work of Austrian-born Columbia University biochemist Erwin Chargaff (1905–2002).[10] Crick was the pattern to Watson's process. Together, with much head scratching and 'help' from others, most notably the English biophysicist Rosalind Franklin (1920–1958) and her spectacular X-ray diffraction images of DNA crystals,* Watson and Crick produced a 3D model of DNA (eventually out of machine-shopped metal) that was remarkable in its elegance (Fig. 2). Confident it was correct, they wasted no time celebrating in the Eagle, a pub where Crick apparently announced to anyone who would listen that he and Watson 'had found the secret of life'.[11]

Watson and Crick's publication was short and sweet, but came with a fully loaded conclusion: 'It has not escaped our notice that the specific pairing we have postulated immediately suggests a possible copying mechanism for the genetic material'.[12] By 'specific pairing' they were referring to the interactions between the two strands of the double helix. Each step in the DNA ladder has three components, a phosphate on the outside, a base on the inside, and a sugar connecting the two—together they make up a 'nucleotide'. The phosphates link adjacent nucleotides together to form the vertical backbone of the

* Franklin's X-ray data were given to Watson and Crick without permission by her King's College London colleague Maurice Wilkins. Franklin was at the time close to solving the structure herself, but was against speculative model building before enough experimental data had been collected. Sadly she died of cancer at the age of 37 and never received proper credit for her critical role in the elucidation of the structure of DNA. Watson, Crick, and Wilkins received the Nobel Prize in Physiology or Medicine in 1962 for the discovery.

ladder, while the bases form the horizontal rungs: base A pairs with base T, and C pairs with G. These so-called Watson–Crick chemical bonds are key: if you know the sequence of bases on one strand of DNA then you know the sequence of its partner strand.

Could it really be that simple? Could this explain how DNA replicates itself in the cell? Yes and no. In what has been described as 'the most beautiful experiment in biology',[13] Meselson and Stahl showed in 1958 that DNA replicates in a semi-conservative fashion.[14] The DNA duplex unwinds and each strand serves as a template for the addition of a complementary nucleotide across from it. The presence of a T residue gives rise to an A, an A yields T, and so on. If both strands of an unzipped helix are replicated in this fashion, the end result is two double helical molecules identical to the original. DNA is emphatically not, however, capable of doing any of this on its own. It is dependent on a battery of proteins to unwind the double helix and add the correct sister bases one by one. DNA may be the secret of life but on its own it is utterly lifeless.

The final piece of the DNA puzzle was to crack the genetic code, a daunting task if ever there was one. It took scores of researchers more than a decade to finish the job and it was Crick, with his fierce intelligence and knack for generating testable hypotheses, who played the role of godfather. On paper the problem was simple cryptography. A protein molecule is a linear chain of amino acids that folds up into a complex three-dimensional structure—the precise sequence of amino acids along the chain is what determines a protein's shape, which in turn determines its biological function. Given that there are four bases in DNA and 20 different kinds of amino acids, a strict one-to-one correspondence was impossible, as was a two-base code, which could only specify $4 \times 4 = 16$ amino acid residues. A three-character code was thus invoked: $4 \times 4 \times 4 = 64$ different three-base combinations, enough to account for all 20 amino acids with room to spare.

And so it proved to be—a non-overlapping triplet code. The South African biologist Sydney Brenner was an active participant in its deciphering, proposing that the DNA triplets be referred to as 'codons'. Figuring out how codons are actually used by the cell to synthesize protein was the hard part. Brenner, Crick, François Jacob (1920–2013), and others had reasoned that in order for genetic information to 'flow', something had to act as an intermediary between DNA, which in eukaryotes resides in the nucleus, and proteins, which are made in the cytoplasm. That 'something' turned out to be RNA (ribonucleic acid), a close chemical relative of DNA found in all cells and in a variety of different forms, some highly abundant and omnipresent, others vanishingly rare and fleeting. RNA differs from DNA in that thymine is replaced by uracil (U), and in 1961 Marshall Nirenberg (1927–2010) and Heinrich Matthaei provided experimental proof that in the bacterium *E. coli* the RNA triplet UUU specifies phenylalanine,[15] the first of the 20 amino acids to be assigned a codon. This would prove to be the case for other organisms as well, prokaryotes and eukaryotes alike. By decade's end, the path from DNA to protein was reasonably clear. DNA acts as a template for the synthesis of RNA using the same principles of base complementarity that underlie DNA replication; it is the resulting single strand of 'messenger' RNA (mRNA)—and the string of codons it contains—that serves as the instructions for the production of protein. Within the confines of the ribosome, the cell's protein manufacturing facility, each of the 20 different types of amino acid is brought into position by one of its very own 'transfer' RNA molecules, whose 'anti-codons' can base pair with the appropriate codons in the mRNA. Simply put, individual amino acids come in one side of the ribosome and a precisely ordered chain of amino acids comes out the other, ready to fold up and fulfil its cellular destiny. Even the simplest prokaryotic cells have thousands of ribosomes floating around, each churning out protein chains at more than 15 amino acids per second.

Having moved from physics to biology in the 1950s, Francis Crick again changed focus (and continents) in the mid-1970s, this time to the study of neuroscience and human consciousness at California's Salk Institute. But he left an indelible stamp on genetic code research in the form of his 'central dogma' of molecular biology—first introduced in

1958, revisited in 1970, and still taught today—which describes the unidirectional flow of genetic information in living organisms. In essence, DNA is transcribed into RNA, which is translated into protein and 'once "information" has passed into protein it cannot get out again'.[16] Half a century on, nothing we've learned invalidates the central dogma as a useful guiding principle.[17] And what has become clear is that the code is essentially universal: the same 64-codon genetic code underlies protein synthesis in all life forms.[18] The French molecular biologist Jacques Monod (1910–1976) said it best: 'what is true for *E. coli* is also true for the elephant'.

Molecules as clocks

Elephants and *E. coli* share a good deal more than their genetic code. Many of the proteins that underlie core cellular processes, including energy generation and protein synthesis, are conserved across vast evolutionary distances. By 'conserved' I mean that the sequence of amino acids in protein X from organism Y is similar, and in some cases virtually identical, to its counterpart in organism Z. The observed degree of similarity depends on the protein and the organisms being compared. A particularly striking example is 'heat shock' proteins (HSPs), which are molecules that play an important role in helping cells deal with environmental stresses. Aligned in a linear fashion amino acid by amino acid, some HSPs in the African elephant, *Loxodonta africana*, are more than 50 per cent identical in sequence to HSPs in the lowly bacterium *E. coli*. Consider too cytochrome *c*, a protein at the heart of cellular respiration in mitochondria. The human cytochrome *c* protein is 105 amino acids long, and 97 of these are the same in an elephant—the human and elephant cytochrome *c* proteins are 92 per cent identical. Your DNA-binding protein histone H4 is 98 per cent identical to the histone H4 protein in a pea plant! This knowledge probably won't leave you feeling any 'closer' to elephants and peas than you did before, but there you have it: it's some of the strongest evidence scientists have for the unity of life on Earth.

A protein's sequence and structure determine its function. Given the extent to which evolutionarily distinct organisms can have proteins with similar amino acid sequences, it is perhaps not surprising that a certain degree of 'mixing and matching' is tolerated. Indeed, one of the ways scientists learn about proteins is to introduce them into 'model' organisms such as *E. coli* and the simple eukaryote yeast. In labs the world over *E. coli* and yeast are routinely engineered to contain genes from organisms that, for experimental and/or ethical reasons, cannot (or should not) be monkeyed with. The foreign gene sequences can be modified at will, and the resulting proteins assessed for their ability to serve as functional replacements for their endogenous counterparts. These so-called structure-function studies have been the lifeblood of biomedical research for decades, contributing greatly to our understanding of how proteins work at the biochemical level. Genetic engineering also underlies much of the biotechnology industry in which life forms are mixed, matched, and modified in order to make proteins and other molecules of potential use to humanity. This includes using *E. coli* as a 'factory' for the production of pharmaceutical agents such as human insulin and growth hormone; improving the yield and nutritional content of crops through genetic modification; and bioremediation, which increasingly involves enhancing the existing biochemical capabilities of microbes so that they can remove pollutants from the environment, clean up spills, and so on. All this sounds very unnatural, and it is. But as will become abundantly clear nature routinely 'experiments' with the genomes of organisms across species boundaries as well, particularly microorganisms. Evolution is not an engineer but it is a tinkerer.

As striking and useful as they are, protein sequence similarities between distantly related organisms tell only one side of the story. What are we to make of the differences? Crick proposed what could be made of them in 1958:

Biologists should realize that before long we shall have a subject which might be called 'protein taxonomy'—the study of amino acid sequences of the proteins of an organism and the comparison of them between species. It can be argued that these sequences are the most delicate expression possible of the phenotype of an organism and that vast amounts of evolutionary information may be hidden away within them.[19]

The idea that molecules might somehow be useful for the study of evolution goes back a surprisingly long way, at least as far as the late 1800s to German zoologist Ernst Haeckel (1834–1919) and his mystical musings about the physical nature of 'Monera' (what we would today call prokaryotes). Yet even with the advent of molecular biology and the realization that DNA is a repository for linear sequence information, little could be accomplished without data. Someone had to figure out how to collect it.

By the mid-1960s, with the code-crackers working overtime, it was still not possible to determine the precise order of bases along a DNA or RNA molecule. But the sequence of amino acids in a protein could, with great effort, be obtained. It was the quiet and unassuming Cambridge biochemist Frederick Sanger, recipient of the Nobel Prize in Chemistry not once but twice, who in the early 1950s first showed it could be done. Sanger chose to hammer away at the two amino acid chains of insulin, a protein of obvious medical importance and one of the very few proteins available in pure form at the time. Using various biochemical procedures, he broke the proteins into small fragments and determined the precise residue present on the end of each fragment. By repeating the procedure, he was eventually able to piece together the complete sequences of the A and B chains, a grand total of 21 and 30 amino acids, respectively.[20] (Sanger worked with cow insulin, but the human and cow proteins proved to be of the same length and differ at only three amino acid positions.) It was the first definitive demonstration that proteins differ not only in amino acid composition, but that different proteins have their own, unique primary amino acid sequence. It was also the first trickle of what would soon become a wave, and eventually a flood, of sequence information available to biologists for study.

Among those first to embrace the idea that deep insight into the origin and evolution of life could be gleaned from molecules were Linus Pauling (1901–1994), Émile Zuckerkandl (1922–2013), Margaret Dayhoff (1925–1983), Walter Fitch (1929–2011), and Emanuel Margoliash (1920–2008), all based in the United States. Dayhoff in particular played a key role in shepherding in the field of bioinformatics, the use of computers and statistics to study sequence data. In the early 1960s, Zuckerkandl and Pauling carried out a landmark analysis of vertebrate haemoglobin,[21] the all-important oxygen-transporting protein in our red blood cells. Together with the cytochrome c-based studies of Fitch and Margoliash,[22] the haemoglobin data supported the idea that protein sequences diverge from one another at a constant rate: the more distantly related two species were—as inferred from the fossil record, for example—the more differences one would see in their amino acid sequences. The 'molecular clock hypothesis' of Zuckerkandl and Pauling was a remarkable and potentially transformative idea. If true—if molecules really could reveal how present-day life forms relate to one another—it opened up exciting new possibilities for the study of evolution. It was the genesis of what today is called molecular phylogenetics: the study of molecular differences to gain insight into the evolutionary relationships among organisms. In principle no organism, no matter how small or simple, need be excluded from consideration for lack of fossil or phenotype.

Let's step back and consider, by way of analogy, the source of molecular sequence variation from species to species. Think about the following make-work exercise. It begins with you convincing nine of your friends to buy this book, and all ten of you writing it out by hand, cover to cover. Each of you then passes her/his hand-written copy on to another ten friends, who do the same. After a few iterations, would it be possible for me to accurately predict who had passed their texts on to whom by comparing all the hand-written copies to one another and to the original version of the book? If none of the scribes made any errors in writing out the text, absolutely no errors whatsoever, then the answer is no. But what are the chances of that? Even if all the participants have impeccable handwriting, everyone will make at least a handful of mistakes: typographical errors, substitutions and deletions of words or phrases, and so on. In 'round 1', it is highly unlikely that any of the mistakes would be the same: each of the first ten versions of the text, having been independently copied out from the original, would contain its own unique set of errors. Texts created in 'round 2' would also have their own errors but—and this is the key—they would contain the same errors as those made by the scribe whose text theirs was copied from. And so on. Once made, errors are 'inherited' faithfully down the line (assuming no one takes it upon themselves to correct any of the errors they see). So by comparing patterns of similarities and differences between all of the handwritten books, it would in fact be possible to tell with reasonable certainty the history of the documents—whom had passed text to whom.*

In the context of molecular phylogenetics, the hypothetical copying errors made by you and the other scribes are analogous to mutations—changes in the DNA sequence (this isn't just a silly plug for my book—manuscript evolution is a legitimate field of scientific investigation[23]). As befitting a molecule whose job it is to store hereditary information, the process of DNA replication within living cells is efficient and highly accurate. But it is far from perfect. Very occasionally, on the order of once per 100,000 additions, the wrong nucleotide is incorporated: a C gets added instead of a T, for example. If the mistake is not caught and corrected (and cells have proteins dedicated to this task), the error will be passed on to daughter DNA strands during subsequent rounds of replication. Mutations are for the most part random—each nucleotide position in an organism's genome is equally likely to be incorrectly copied—and because there are

---

* This seems like a rather good idea. I'm willing to experiment if you are, but I suggest starting with 99 friends just to be sure.

more codons in the genetic code than there are amino acids in proteins, many DNA mutations do not result in a change in protein sequence. But some do, and such mutations have the potential to influence the structure and function of the protein, and the fitness of the organism possessing it.

Darwin in the details

'The time will come I believe, though I shall not live to see it, when we shall have fairly true genealogical trees of each great kingdom of Nature'.[24] So wrote Charles Darwin (1809–1882) to his fellow Englishman and scientific supporter Thomas Huxley (1825–1895). The year was 1857, two years before On the Origin of Species was published. Darwin's world was not our world; it did not include DNA or genetics or any real notion of the mechanisms of heredity as presently understood. And Darwin's knowledge of the diversity of life was limited by today's standards, particularly (and not surprisingly) with respect to the microbial biosphere. But Darwin was a keen observer, a deep-thinking naturalist who spent his life trying to make sense of the organisms he could see with his own eyes—beetles, birds, and barnacles, Great Danes and miniature Chihuahuas, mice and men—and the endless variation exhibited by such creatures. Piecing together his own findings with evidence and ideas from diverse areas of science, including geology, comparative anatomy, and 'population theory', Darwin formulated the concept of evolution by natural selection whereby life forms that are better adapted to their environment are more likely to survive and produce offspring, which, resembling their parents, are more likely to bear these same adaptive traits.* Darwin's thesis was something to be tested and, if possible, falsified using the

---

* Darwin was influenced by the writings of Thomas Robert Malthus (1766–1834), the British scholar and author of a famous and controversial book entitled An Essay on the Principle of Population. Of particular interest to Darwin was Malthus's point that organisms tend to produce far more offspring than will survive in a world with finite resources.

scientific method. It is now a theory in the formal scientific sense of the word, a unified body of knowledge no different than Newton's law of gravity, confirmed by 150-plus years of observation and experiment. The notion that evolution is 'just a theory' does not compute.

This is not to say that scientists don't debate evolution. They certainly do. But the debate revolves around how it happens and what it means, not whether it occurs. A product of the nineteenth century, gaps in Darwin's theory were gradually filled during the twentieth in response to new information from genetics and molecular biology about the patterns and processes of inheritance. Mutations in DNA were understood to be the ultimate source of phenotypic variation upon which natural selection can act. Microorganisms were shown to be no different than any other form of life—they have DNA and proteins and phenotypes that impact their goodness-of-fit with the environment and, consequently, their reproductive success. Viruses may not be alive but they too evolve; the signature of natural selection can be seen in their genes. Darwin's general principles of evolution are more relevant to modern biology than ever before.

And yet it has become increasingly difficult to square Darwin's vision of the tree-like diversification of life with the revelations of molecular phylogenetics and genomics. Darwin's 'genealogical trees' were but twigs on a single grand bifurcating tree of life, a tree that once resolved would depict the relationships within and between 'each great kingdom' past and present. It's a wonderful vision, one that has inspired generations of biologists to work towards its fulfilment, myself included. But the tree of life has come upon hard times, for two reasons. The first is that prokaryotes do not rigorously obey the rules of inheritance that govern the evolution of multicellular organisms. To be sure, prokaryotic cells divide and their daughter cells divide and their descendant populations gradually diverge from one another in a 'vertical' fashion, just as animal and plant species do. But they also engage in what is called horizontal (or lateral) gene transfer. We've known that prokaryotic cells can exchange DNA with one another since the early 1950s—prokaryotes have 'sex' too—but no one could have predicted the extent of their promiscuity. Analyses of prokaryotic genomes show them to be complex mosaics of genes taken from here, there, and everywhere, often with blatant disregard for species boundaries. All this mixing and matching has led researchers to vigorously debate the relative importance of vertical and horizontal gene flow in prokaryotic evolution, and to question whether prokaryotic species even exist. The media, of course, love a good row. In 2009 *New Scientist* magazine did its part to stir up controversy by running the following cover: 'Darwin was wrong: Cutting down the tree of life'.

The second reason why a tree is an inaccurate representation of the diversification of life is the subject of this book: endosymbiosis. The endosymbiotic origins of mitochondria and chloroplasts are instances in which nature's mixing and matching involved not just genes but *entire single-celled organisms*. In the case of mitochondria the evidence suggests that the evolutionary transition from endosymbiotic bacterium to energy-converting organelle was a singular, exceptional event, one that ultimately spawned complex life in the form of the eukaryotic cell. Chloroplasts appear to have evolved from endosymbiotic cyanobacteria only once in the history of eukaryotic evolution but they have spread horizontally across the tree from eukaryote to eukaryote on numerous occasions, driven by the significant evolutionary advantages that come with acquiring the ability to harness the energy of the sun. There is increasing enthusiasm for the notion that the 'tree of life' is in fact a web of life—a complex net through which genetic information has flowed both vertically and horizontally for more than three billion years.*

Was Darwin wrong? The answer depends on the extent to which one chooses to bring him to task for things he could know nothing about. The molecular biology revolution made it possible to build upon Darwin's basic principles, to open a new window on the

---

* In discussing the significance of horizontal gene transfer I have focused on prokaryotes, but single-celled eukaryotes also exchange DNA in this manner, as do some multicellular groups (e.g. plants).

evolution of life on Earth. One of the first fundamental scientific problems to be tackled using the tools of molecular phylogeny was the endosymbiont hypothesis for the origin of mitochondria and chloroplasts. In the chapters that follow we will trace the threads of endosymbiosis forwards and backwards through deep time, through the eyes of the researchers who wove them into a fundamentally new level of understanding of life's history. Some of these scientists are well known; most are not. The same is true of the organisms they studied. Pond scum? Oceanic phytoplankton? Malaria parasites? The ideas are old but the data are new. Darwin would have been delighted to hear them.

3

THE SEEDS OF SYMBIOSIS

The British evolutionary biologist J. B. S. Haldane (1892–1964) was once asked by a group of theologians what his scientific investigations had revealed about the mind of the Creator. His response? 'God has an inordinate fondness for stars and beetles'.* Biochemists are inordinately fond of lines and circles. In trying to make sense of the flow of energy through the cell's interior, metabolic 'maps' are drawn—diagrams in which molecules make their way round a circle or down a line, transformed along the way by enzymes, proteins dedicated to turning compound A into compound B, often with the assistance of helper compound C. Consider glycolysis, the 'sugar splitting' pathway that lies at the energetic heart of the cell: ten different enzymes participate in a chain of reactions that convert glucose into a chemical called pyruvate. In oxygen-utilizing eukaryotic cells pyruvate is made in the cytosol but is quickly shunted to the mitochondrion where it enters another pathway, the citric acid cycle (also known as the Krebs cycle after its discoverer, Sir Hans Adolf Krebs (1900–1981)). Through the sequential action of eight additional enzymes, each turn of the citric acid cycle produces essential electron-carrying compounds that feed the so-called oxidative phosphorylation pathway. And it is *this* pathway that is the moneymaker: it generates adenosine triphosphate, or ATP, the cell's energy 'currency'. Nothing in life happens without ATP.

---

* Sources are conflicted as to precisely what Haldane said, but there are undoubtedly more than 300,000 species of beetles on Earth. That's a lot of beetles.

Cellular metabolism occurs on a scale and at a pace that is difficult to wrap one's mind around. At any given moment the human body contains a mere 250 grams of ATP; but in order to keep pace with energy demand, we crank out the equivalent of our own body weight in ATP every 24 hours. This astonishing rate of turnover is possible because enzymes can catalyse more than a million reactions per second. It is a testament to human ingenuity and perseverance that researchers have managed to 'write the book' on hundreds of enzymes and their target substrates, to connect the biochemical pathways in which they participate, and to track them back to the genes that encode them. So advanced is the state of twenty-first-century bio-informatics that a fairly accurate picture of the metabolism of an organism can often be obtained simply by perusing its genome: enzyme-encoding gene sequences are identified and the biochemical dots are connected *in silico*, no laboratory experimentation required.

Lines and circles are essential visual aids in our quest to understand the chemistry of life. But they are also misleading in that they give the impression that the interior of the cell is metabolically neat and tidy. In reality nothing could be further from the truth. Despite how they are depicted, biochemical pathways are not sub-cellular road maps along which chemical compounds move systematically from station to station where the appropriate enzymes wait patiently to spring into action. The intracellular milieu is a complex stew brimming with proteins, nucleic acids (DNA and RNA), and a myriad of other molecules—carbohydrates, sugars, fats, etc. The concentration of any particular molecule is typically low, but the total molecular concentration is high—so high that the cytosol is more like a gel than a liquid. It is amazing that anything biochemically 'useful' happens at all. Molecules bounce around like Ping-Pong balls; collisions between enzymes and potential substrates happen; usually nothing comes of them. But if the chemistry is right, reactions will proceed. These are the reactions that sustain life.

Cellular metabolism is a rough-and-tumble world in which chaos rules. Yet from chaos comes opportunity. The universal nature of

Earth's biochemistry is such that organisms as different as beech trees and bacteria can, at the level of molecules, speak the same language. DNA is DNA, carbon dioxide is carbon dioxide, glucose is glucose—it doesn't matter who made it or how. From this perspective it's not difficult to imagine distantly related organisms forging intimate metabolic connections with one another over the fullness of evolutionary time. As we've learned already, ATP synthesis in eukaryotes often involves the exchange of chemical intermediates such as pyruvate between the mitochondrion and the cytosol. In plants and algae, the biochemical processes taking place inside their light-harnessing chloroplasts are similarly linked to those of the eukaryotic 'host'. Mitochondria and chloroplasts are in fact so woven into the biological fabric of their host that it can be difficult to recognize them for what they actually are—or were. The evolutionary 'experiments' that gave rise to them have long since run their course.

So how *do* we know that mitochondria and chloroplasts evolved from bacteria by endosymbiosis? The best way to understand how is to consider a time when we did not. Scientists have pondered the evolutionary significance of the compartmentalized nature of the eukaryotic cell for as long as they have been able to observe it, and before modern biochemistry—and long before genes, genomes, and molecular phylogenetics—a handful of researchers managed to put two and two together. Alas, their ideas were destined for scientific obscurity; they were square pegs in round holes, and no amount of pounding was going to make them fit comfortably within the conceptual framework of the day. More than half a century would pass before they were 'rediscovered' and science caught up.

Living together, evolving as one?

Nature makes strange bedfellows. The word *symbiosis* comes from the Ancient Greek 'together' and 'living', and its use in scientific discourse dates back to the second half of the nineteenth century. It is easy to visualize: picture a snarling crocodile, jaws wide open, with a dainty plover picking food from its jagged teeth with not a care in the world. Or a red-beaked oxpecker bird hitching a ride on the back of a zebra, dutifully plucking ticks from its fur. But the concept of symbiosis is, and always has been, a magnet for controversy. Once the biological partners in a symbiotic relationship have been identified, the question immediately becomes 'what's in it for whom'?

If both partners benefit—as in the case of an oxpecker enjoying a free lunch (in the form of tasty bugs) and the mammal it rides getting picked clean of parasites—then the symbiosis is considered to be mutualistic. If, on the other hand, one of the partners is harmed as a result of the interaction then it is, technically speaking, parasitism. Symbiotic relationships can be obligate, meaning that the organisms cannot live without one another, or facultative, whereby one or both partners can get by in the event of a break-up. They can be ectosymbiotic—when one organism lives *on* another organism—or endosymbiotic, which involves an organism living *within* another. The definitions are straightforward, but in nature the boundaries between these different forms of symbiosis are often blurred.

The concept of symbiosis stems largely from the work of three biologists, Simon Schwendener (1829–1919) of Switzerland, and the Germans Anton de Bary (1831–1888) and Albert Frank (1839–1900). And it began with consideration of the most unassuming of biological specimens—lichens [like-ens]. In 1867 Schwendener proposed that lichens are composite organisms comprised of a fungus (a 'myco-biont') and an alga (a 'photobiont').[1] Lichens are, it must be said, spectacularly easy to overlook. They exist as pale green or rust-coloured mats, often with a scaly, crusty appearance; they will grow (at a snail's pace) on just about any stable surface that gets a bit of sunlight and at least occasionally gets wet—rocks, tree trunks, hard soil, brick walls, roofs, and the like. They don't have leaves and they don't sway in the breeze, but lichens have traditionally fallen within the domain of botany, mainly because botanists were the only ones who paid them any notice.

The 'body' of the lichen is the thallus, a multi-layered network of fungal filaments with algal cells embedded within it.* The algae do what photosynthetic organisms are wont to do—make carbohydrates from sunlight, water, and carbon dioxide. The fungus takes its share of this handy source of energy and in return provides the alga with nutrients, protection from desiccation, and a broad surface on which to spread out in search of photons. In this sense the lichen thallus is not unlike the leaf of a plant or tree. It's a strange but popular marriage of convenience; modern research shows that fungal–algal partnerships of this nature have evolved time and time again over the past 500 million years.

Schwendener's 'dual hypothesis' of lichens did not sit well with contemporary experts. By the time he began his studies, hundreds of lichen species had already been described, the general morphology of which did not remotely resemble any known alga or fungus. But Schwendener was an expert light microscopist, and his knowledge of such organisms living naturally apart from one another allowed him to probe the fine structure of lichens and see what most lichenologists could not see—or did not want to see. Resistance to the idea of symbiosis was deep-seated; it ran counter to nineteenth-century views on the autonomous nature of the organism. In trying to contextualize what lichens are and what they are not, biologists were navigating uncharted territory. *Are* lichens organisms? If so, how should they be classified? Schwendener's dual hypothesis was, in the words of Canadian historian of biology Jan Sapp, 'an abomination to systematists'.[2]

Speculation and philosophical debate eventually yielded to hard data. Schwendener and his colleagues were able to show that the algal and fungal components of at least some lichens could be teased apart and put back together again in the lab, thereby demonstrating beyond all shadow of a doubt that lichens are indeed 'double organisms'. But while Schwendener was proven correct, his views on the

---

* Lichens with cyanobacterial 'photobionts' have also been described. Occasionally algae and cyanobacteria are part of the same lichen.

precise nature of the symbiotic partnership were controversial. He insisted that the relationship between fungus and photobiont was one of master and slave—the fungi were deemed to be the ones in control. The German de Bary was less anthropomorphic and more flexible. He defined symbiosis as 'the living together of unlike organisms'.[3] He envisioned lichen symbioses as representing a continuum from mutualism through to full-on parasitism, depending on the specifics of the fungi and algae involved and how long they had been eking out an existence in one another's company.

More generally, de Bary, Frank, and others argued that lichens were not one-off freaks of nature, but were in fact manifestations of an important biological phenomenon that needed to be taken seriously. It wasn't idle speculation. In 1885, Frank, who began using the term symbiosis around the same time as de Bary, described another important and widespread symbiosis involving fungi, in this case with the roots of trees. 'Mycorrhizal' fungi serve as conduits for the uptake of essential nutrients into the tree's roots. The pay-off for the fungus, of course, comes in the form of sugar derived from photosynthesis. It is difficult to overstate the importance of these so-called root fungi. Upwards of 90 per cent of 'higher' plants are known to engage in mycorrhizal symbioses, and it is thought that green algae probably could never have colonized the land half a billion years ago—and given rise to plants and trees—had fungi not been there to 'help'.

By the end of the nineteenth century symbiosis had become, if not mainstream, rather more palatable to the average biologist. It was inevitable. The more scientists learned about the natural world, the more examples of intimate relationships between organisms of all shapes and sizes were discovered, and the more scientists had to deal with the implications. Although few would have believed it at the time, these implications extended far beyond the conundrum of how to classify composite organisms. The concept of symbiosis had the potential to wreak havoc with traditional views of Darwinian evolution; those who sought to account for it were inspired to think differently about how life forms interacted with one another and diversified over time.

Little green slaves

Let us imagine a palm tree, growing peacefully near a spring, and a lion, hiding in the brush nearby, all of its muscles taut, with bloodthirsty eyes, prepared to jump upon an antelope and to strangle it. The symbiotic theory, and it alone, lays bare the deepest mysteries of this scene... The palm tree behaves so peacefully, so passively, because it is a symbiosis, because it contains a plethora of little workers, green slaves (chromatophores) that work for it and nourish it. The lion must nourish itself.

These are the words of the Russian biologist Constantin Mereschkowsky, and have been quoted and discussed many, many times in the scientific literature. They certainly are eloquent, but why do biologists find them so fascinating? They were published in 1905.[4] Mereschkowsky (1855–1921) is credited with having articulated the very first version of the endosymbiont hypothesis for the origin of eukaryotic organelles, specifically chloroplasts. He too was inspired by lichens, but only after having devoted much time and effort to the exploration of aquatic organisms. Of particular fascination to Mereschkowsky were diatoms, ubiquitous single-celled algae often referred to as the 'jewels of the ocean' because of their ornate silica cell walls and lustrous golden colour. As a means of addressing fundamental questions in biology, Mereschkowsky recognized the immense value of looking for answers inside the cell. In the early 1900s, having established himself at Kazan University in Russia, he contributed to a reorganization of diatom taxonomy based on sub-cellular structure, in particular the features of their pigmented bodies, or 'chromatophores'—chloroplasts. Combining his knowledge of plant and algal cell biology with ideas about symbiosis, Mereschkowsky developed the concept of *symbiogenesis*, which he defined as: 'the origin of organisms through the combination and unification of two or many beings entering into symbiosis'.[5] From distinct forms of life living in close association for extended periods of time, he believed that entirely new organisms could evolve.

Mereschkowsky was not alone in thinking symbiotically about the evolution of life. In the late 1800s, for example, the German biologist Andreas Schimper (1856–1901), a former student of the lichenologist Anton de Bary, had raised the possibility that chloroplasts might be of endosymbiotic origin:

> If it can be conclusively confirmed that plastids [chloroplasts] do not arise de novo in egg cells, the relationship between plastids and the organisms within which they are contained would be somewhat reminiscent of a symbiosis. Green plants may in fact owe their origin to the unification of a colourless organism with one uniformly tinged with chlorophyll [photosynthetic pigment].*

It's one of the more significant footnotes in the history of biology, and it was literally that—a footnote in a paper Schimper published in 1883,[6] long before Mereschkowsky began writing on the subject.

What made Mereschkowsky's initial contribution to the field unique was the extent to which he assembled evidence from distinct lines of inquiry into a robust case for the endosymbiotic origin of chloroplasts. First, and quite naturally, Mereschkowsky appealed to the increasingly well-supported phenomenon of symbiosis—that organisms could live and evolve together didn't need to be imagined, it could be observed. He specifically discussed instances in which colourless amoebae were known to have 'zoochlorellae' (green algae) living inside them. The only difference between zoochlorellae and chloroplasts, Mereschkowsky argued, was that the former could live and divide outside its host, whereas the latter apparently could not.

Second, Mereschkowsky pointed to the works of the Swiss botanist Carl Wilhelm von Nägeli (1817–1891), Schimper, and others who had demonstrated the continuity of chloroplasts. Within the cells of plants

---

* Biologists often use the terms 'chloroplast' and 'plastid' interchangeably. The distinction between them needn't concern us. For consistency and simplicity, I have chosen to use the term chloroplast throughout this book.

and algae, chloroplasts are not generated 'from scratch' each time the main cell divides. Chloroplasts themselves divide; they come from pre-existing chloroplasts, which are passed on to daughter cells upon division.

Finally, Mereschkowsky went to great lengths to demonstrate that 'there are organisms that we can regard as free-living chromatophores'. He compared what was known about the physiology and structure of chloroplasts with the attributes of 'cyanophyceae'— prokaryotic organisms that are today called cyanobacteria. He considered the similarities between the two to be 'great and obvious': chloroplasts and cyanobacteria are both small, round, blue-green pigmented bodies that lack a nucleus; both assimilate carbon dioxide; and both proliferate by division.[7]

Mereschkowsky's 1905 paper was a masterwork of logic. Over the next 15 years he followed it up with a string of well-reasoned, increasingly exhaustive, and evidence-rich articles; what he considered his pièce de résistance was published in French in 1920, the year before he died.[8] During this period Mereschkowsky's name became synonymous with the idea that chloroplasts had evolved by endosymbiosis, and today he is often hailed as the 'founding father' of symbiotic theory. Against the backdrop of 150 years of symbiosis research his story is often portrayed as one of neglect and rediscovery. But we must remind ourselves again that Mereschkowsky wasn't the only player in the game.

The plant physiologist Andrey Famintsyn (1835–1918) also made substantial contributions to symbiotic theory. Famintsyn, a powerful figure in Russian biology, was an experimentalist through and through. He insisted that the best way to prove that chloroplasts were endosymbiotically derived was to show that they could be cultured on their own, outside the confines of the cell in which they lived. Predictably, Famintsyn did not approve of the more free-wheeling observational-theoretical approach taken by his junior 'colleague' Mereschkowsky. The two were, for all intents and purposes, competitors—they carried out their work independently, each acknowledging the other only rarely. Mereschkowsky, among others, felt that Famintsyn was misguided in thinking that definitive proof of symbiogenesis must come from the separation and cultivation of the various subcomponents of the cell. Famintsyn had been successful in culturing algal endosymbionts from various amoebae and inverte-brates, which gave him hope that he might also succeed in the case of plant and algal chloroplasts. But he did not succeed, and the reason seemed to be obvious to everyone but himself. If the process of symbiogenesis results in the evolution of a single new organism, then growth of its constituent parts in isolation should be impossible. The more ancient the symbiogenetic event in question, the more likely such experiments were to fail.

And what of Mereschkowsky's lions and palm trees?

> Let us imagine each cell of the lion filled with chromatophores [chloro-plasts], and I have no doubt that it would immediately lie down peace-fully next to the palm, feeling full, or needing at most some water with mineral salts.[9]

It's a beautiful piece of imagery. The irony, though, is that the lion gets its energy from mitochondria, which Mereschkowsky steadfastly dis-missed as being of endosymbiotic origin. But he was quite right about a solar-powered lion wanting to lie down. In fact, as pointed out by Nick Lane of University College London, a lion filled with chloroplasts would have no choice—without the surface area of a tree to capture enough sunlight, it simply wouldn't have the energy to stand.

The quirky side of evolution

Famintsyn was unsuccessful in his attempts to cultivate chloroplasts, but at least he never claimed to have done so. This was the unfortunate situation that the French biologist Paul Portier (1866–1962) found himself in during the course of his investigations into the nature of mitochondria. Portier didn't think that mitochondria had evolved from bacteria; he believed that mitochondria *were* bacteria. And he announced to the world that he had succeeded in culturing them from diverse organisms, including plants, insects, and even animal tissue.

Portier was a known quantity in the scientific circles of Europe in the early 1900s. He was connected to (and was at times supported by) none other than Albert I, Prince of Monaco (1848–1922), an enthusiast and early backer of oceanographic exploration. Aboard one of Prince Albert's ships, Portier carried out important research that contributed to the discovery of anaphylaxis, work that resulted in a Nobel Prize for his mentor Charles Richet (1850–1935) (but not for Portier himself).* As a young man Portier had received a medical degree but ultimately decided to pursue a career in research. His interests were exceptionally broad, ranging from microbiology to entomology, typically from the perspective of comparative physiology.

In 1918 Portier published a book (dedicated to Prince Albert) entitled *Les Symbiotes*,[10] the book for which he is now infamous. In a nutshell, he proposed that complex cells were dependent on bacterial symbionts—'symbiotes'—for life. Portier's hypothesis was bold and far-reaching, in keeping with his scientific curiosities and penchant for thinking big. His symbiotes were deemed to be present in plants, animals, and all manner of unicellular eukaryotes, and they were said to play an important role in, well, basically everything: metabolism, development (in multicellular organisms), speciation, cancer; they might even have been the very seeds from which life on Earth had first sprung. In the words of Jan Sapp, 'few biological problems were left untouched by Portier's theory'.[11] An anonymous 1919 review of Portier's book published in the British journal *Nature* described it as 'a lively exposition of heresy'.[12]

---

* To commemorate the fiftieth anniversary of the discovery, the faces of Portier, Richet, and Prince Albert appeared on a Principality of Monaco postage stamp in 1952, along with a picture of the Prince's yacht *Princesse Alice II*. The stamp also included a picture of the Portuguese man-of-war *Physalia*, the bizarre colonial marine invertebrate from which a deadly toxin was isolated and used to study the nature of allergic reactions.

Les Symbiotes caused a brouhaha in France. The fallout is best summed up by consideration of a follow-up book published a year after Portier's by his fellow countryman Auguste Lumière (1862–1954; together with his brother Louis, Lumière invented the cinematograph). The title of Lumière's book was *Le Myth des Symbiotes*.[13] Ouch! In his critique of Portier's work, Lumière paid lip service to the increasingly recognized role of microorganisms in diverse biological phenomena. But on the whole his views reflected the Pasteurcentric sentiments of the day, among bacteriologists at least: bacteria were primarily agents of disease, they did *not* play a role in 'normal' cell biology and animal physiology. Lumière questioned the global significance of symbiosis and Portier's justification for pushing his 'theory' to such seemingly absurd lengths. Theoretical considerations aside, the implicit message coming from *Le Myth des Symbiotes* was that Portier's claims to have cultured mitochondria were not to be trusted. In the months following publication of *Les Symbiotes*, Portier came under pressure from microbiologists at the Pasteur Institute to demonstrate his procedures and show that his results were valid. The resulting back-and-forth ended in a stalemate, but by then the damage was done: Portier's reputation as a competent experimentalist was damaged and his grand hypothesis was ignored.

With the benefit of hindsight it is reasonable to conclude that in attempting to extract and cultivate mitochondria from healthy animal tissue, Portier had simply gone beyond the limits of sterile laboratory technique. The microorganisms he cultured were in all likelihood contaminant bacteria. To his credit, Portier acknowledged this possibility in his debates with the 'Pasteurians', and had always believed that it would be difficult to culture mitochondria because of their presumed adaptation to intracellular life. But the problem was that the precise relationship between les symbiotes and mitochondria was never entirely clear. Portier's symbiotes were capable of exhibiting great variation in form and function, so much so that the central tenets of his hypothesis were essentially unfalsifiable using the scientific tools available at the time. It is open to interpretation what Portier actually believed he had cultured. A follow-up to *Les Symbiotes* was drafted but never published; Portier remained active in science well into his eighties, but never again worked specifically on symbiosis or 'mitochondria'.

Meanwhile, on the other side of the Atlantic, an anatomy professor at the University of Colorado by the name of Ivan Wallin (1883–1969) fell victim to the same experimental trap. Like many biologists at the time, including some of the Russian 'symbiogeneticists', Wallin questioned the extent to which Darwinian natural selection was powerful enough to cause large-scale changes in the biology of organisms. He began publishing on the role of symbiosis in evolution and development in the early 1920s; in 1927 he spelled out his most significant ideas in a book entitled '*Symbionticism and the Origin of Species*'.[14] Wallin believed that speciation, an important, unsolved problem in evolutionary biology, was triggered by symbiotic bacteria altering the heredity patterns of the 'higher organisms' with which they had become intimate: 'The simplest and most readily conceivable mechanism by which the alteration takes place would be the addition of new genes to the chromosomes [of the "higher organism"] from the bacterial symbiont'.[15] With each speciation event, Wallin proposed, mitochondria evolved anew from freshly acquired symbionts. We now know that this is not what happened: molecular sequence data tell us very clearly that mitochondria arose but once in the history of life. But in thinking about the problem Wallin appears to have been the first to explicitly consider the movement of genetic material from endosymbiont to host, a phenomenon that today is recognized as an important step in the melding that takes place early in the evolution of an organelle by endosymbiosis. We shall return to this issue in Chapter 7.

Like Portier, Wallin believed that it was possible, and important, to demonstrate that mitochondria could be cultured. He was aware of the problems that had plagued Portier and was hell-bent on avoiding them—he went to great lengths to convince his readers of the rigorous nature of his culturing experiments. Alas, like Portier, his claims to have successfully propagated mitochondria *in vitro* were ridiculed, and his novel ideas about bacteria, mitochondria, and speciation were conveniently dismissed.[16] According to Donna Mehos, 'Wallin's critics simply didn't like what he was saying, and they used the most powerful weapon available in that period of biology—methodological criticism'.[17] Isolated and scorned, Wallin's research career was finished less than a decade after it began.

Harvard palaeontologist Stephen J. Gould (1941–2002) once referred to symbiosis as the 'quirky and incidental side' of evolution.[18] Considering the ideas of Portier and Wallin one might be tempted to agree with him. But Gould's reasons for saying so had nothing to do with the wild speculations of these early theorists; he most likely had never heard of them. If he wrote about microbes at all (which he did only rarely) it was seemingly to lament the fact that for the first half of the history of life on Earth prokaryotes took their sweet old time evolving into eukaryotes, from which the more interesting multicellular animals ultimately evolved.

Gould himself had a reputation as a somewhat unorthodox thinker, having proposed, together with Niles Eldridge, the concept of 'punctuated equilibrium'. In the 1970s, as a possible explanation for the relative paucity of transitional forms seen in the fossil record, Eldridge and Gould hypothesized that organismal evolution proceeds in something of a herky-jerky fashion—long periods of relative stasis are interrupted by 'rapid' bursts of speciation induced by changes in the environment.[19] The hypothesis was viewed by some as an affront to Darwin-style gradualism where evolutionary change is constant and species are transformed into other species in a smooth, steady manner.

Within the broader framework of twentieth-century biology, however, Gould was very much a traditionalist. His views lay firmly under the umbrella of the so-called 'modern evolutionary synthesis' (or neo-Darwinian synthesis), the marriage of Darwin's natural selection with Mendel's pea plant genetics. In the late 1920s—at precisely the same time that Wallin was floating his radical ideas about 'symbionticism'—

American geneticists such as Hermann Muller (1890–1967) were making spectacular advances in elucidating the genetic basis of inheritance. In the laboratory, X-rays were shown to cause seemingly random mutations in fruit flies. These mutations produced phenotypic variations indistinguishable from those observed in natural fly populations—changes in eye colour, wing shape, leg length, and so forth. Such variation was the substrate upon which natural selection could act and now it had a root cause: changes in the structure of the gene. And where were these genes? They were in the chromosomes of the cell nucleus. When present, bacterial symbionts and organelles such as mitochondria and chloroplasts were *not* in the nucleus, they resided in the cytoplasm, and in 1926 the American geneticist Thomas Hunt Morgan (1866–1945) concluded that 'the cytoplasm may be ignored genetically'.[20] Morgan was a hugely influential figure—he would eventually receive a Nobel Prize for helping to elucidate the role of the chromosome in heredity. If Morgan said the cytoplasm could be ignored, that was all that most evolutionary biologists needed to hear.

And so as the modern synthesis took hold of biology in the 1930s and 1940s, symbiosis became increasingly irrelevant as a putative source of evolutionary innovation. Unlike Gould, and many a neo-Darwinist before and after him, Edmund Beecher Wilson (1856–1939) was well aware of the ideas of Mereschkowsky, Portier, and Wallin. He just didn't find them very interesting. Wilson, often referred to as 'America's first cell biologist', was the author of a groundbreaking textbook entitled *The Cell in Development and Heredity*, in which he dismissed the possibility that mitochondria and chloroplasts had evolved from once free-living organisms by endosymbiosis. In the third and final edition of *The Cell*, published in 1925, Mereschkowsky's ideas were considered to be 'entertaining fantasy'; those of Wallin were deemed 'too fantastic for discussion in polite biological discourse'.[21] Wilson was nevertheless clearly intrigued: 'it is within the range of possibility that they may some day call for more serious consideration'. And with that the seeds of symbiosis lay dormant for the better part of

50 years. They were picked up, planted, and nurtured back to life by an ambitious young American biologist with the first name Lynn.

Rebel with a cause

In science, as in life, it is difficult to have a truly original idea. Timing is also everything. Lynn Margulis (1938–2011) was ahead of her time, but only just. To say that she struggled to publish her first scientific paper on the topic of endosymbiosis is a massive understatement—it was rejected 'fifteen or so' times before finally appearing in *Journal of Theoretical Biology* in 1967.[22] At the time she was Lynn Sagan, the 29-year-old wife of astronomer and science popularizer Carl Sagan (1934–1996).* Publication of her first book, *Origin of Eukaryotic Cells*,[23] was no picnic either (after five months of waiting she received a rejection letter from the publishing house, no explanation given). But by the time it appeared in 1970 researchers were increasingly interested in what she had to say.

Margulis was a scientific rebel whose cause was the microbial biosphere. Throughout her career of over 50 years she railed against convention in biology. She believed in the importance of a holistic approach to the study of life on Earth, and she 'walked the talk' at a pace few scientists could match. She was a long-time collaborator and supporter of British chemist-environmentalist James Lovelock and his so-called 'Gaia hypothesis', the notion that the Earth and all its organic and inorganic components comprise a single ecosystem. Margulis referred to Gaia as a 'tough bitch', not a single organism but a self-regulating entity that would go on controlling itself long after we humans disappear.[24] And the microbes, she argued, would be the ones in the drivers seat, as they have been for more than three billion years.

---

* Margulis had two children with her first husband, Dorion Sagan and Jeremy Sagan. Lynn and Dorian would publish many scientific articles and best-selling books together. She later married (and divorced) the biochemist Thomas Margulis.

What would be the legacy of that 'mammalian weed' called *Homo sapiens* in the fossil record? 'The squashed remains of the automobile'.

A 1991 profile of Margulis and her controversial ideas ran with the headline 'Science's Unruly Earth Mother'.[25] In it the British evolutionary biologist John Maynard Smith (1920–2004) summarized how many felt at the time:

> Every science needs a Lynn Margulis...I think she's often wrong, but most of the people I know think it's important to have her around, because she's wrong in such fruitful ways. I'm sure she's mistaken about Gaia, too. But I must say, she was crashingly right once, and many of us thought she was wrong then, too.

What Margulis was 'crashingly right' about, of course, was endosymbiosis. She is widely revered as its modern-day champion. If asked who first conceived of the notion that mitochondria and chloroplasts evolved from bacteria, most scientists (professional biologists included) wouldn't hesitate—it was Margulis.* Yet as we have seen, the raw ingredients for a comprehensive theory of evolution by endosymbiosis can be found scattered across more than a century of science published in German, Russian, French, and English. Margulis formulated her own recipe. In the 1960s and 1970s her knowledge of the early literature on symbiosis was, by her own admission, 'woefully limited'; at the time she believed her ideas to be 'entirely original'.[26] And in many ways they were. Margulis had to fight tooth and nail to convince her peers that endosymbiosis had played a role in the evolution of complex cells. But all things considered, it was a battle whose time had come—and she was the right person to lead the charge.

Born Lynn Petra Alexander, Margulis grew up the oldest of four sisters, city-smart and fearless in South Side Chicago. Vivacious, precocious, and impatient, she attended the University of Chicago at the age of 16 because 'they let me in'. It was as a Master's student at the

---

* I confess that it is anecdotal 'evidence' that leads me to this conclusion, but I would bet my lunch on the results of a properly conducted scientific experiment.

University of Wisconsin that Margulis first became aware of 'non-Mendelian' inheritance, a curious phenomenon in which cytoplasmic (extra-nuclear) 'genetic factors' could influence the passage of an organism's traits from one generation to the next. At the time nobody knew what these 'factors' were; most geneticists were content to ignore them. One of her professors, the pioneering cell biologist Hans Ris (1914–2004), used to read aloud to his advanced cytology class from Wilson's *The Cell*—it was in this context that she caught her first, faint whiff of the 'radical' ideas of Mereschkowsky and Wallin. The late 1950s were exciting times for biology: together with Ris, Margulis's graduate supervisor Walter Plaut (1923–1990) and others were pushing the limits of the electron microscope; the fields of biochemistry, genetics, and cell biology were actively cross-fertilizing one another. The cell's complex interior was coming into focus like never before and Margulis had a front-row seat.

It took time for Margulis to learn to look inside the cell; she was hard-wired to think outside it. In Wisconsin she studied the biology of single-celled amoebae, and with inspiration and support from Ris and Plaut, began to ponder the broader evolutionary significance of their membrane-bound cytoplasmic organelles. As a 22-year-old mother of two Margulis started doctoral research at the University of California (Berkeley) and became increasingly smitten with microbial symbioses. She scoured the literature to keep up with the latest developments, believing that significant advances in evolutionary theory would come not from the study of organisms in isolation, but by considering how they went about their daily lives—as members of dynamic, interacting communities. She was appalled by what she perceived as a vast academic gulf between genetics—as studied by geneticists working head-down in genetics departments—and evolution, which, at Berkeley at least, fell under the domain of palaeontology. Genetics, Margulis believed, held 'the key to evolutionary history'.[27] And she became ever more convinced that Morgan's advice to 'ignore the cytoplasm' was plain wrong. She embarked on what would become a career-long journey off the beaten path, connecting the dots between bacteria, eukaryotic organelles, and the evolution of life, looking small but always thinking big.

Margulis achieved fame as a theorist but cut her scientific teeth as a card-carrying experimentalist. In her very first publication, written with Plaut in 1958, she presented indirect evidence for DNA in the cytoplasm of an amoeba.[28] At UC Berkeley she produced similar data for another microbe, the pond-dwelling photosynthetic alga *Euglena*. In and of themselves, these results weren't going to ruffle too many feathers, but Margulis reckoned she knew precisely what they meant. Others did too. By the time her Berkeley *Euglena* work was published in 1965,[29] Ris and Plaut had obtained electron microscopic evidence for DNA in the chloroplast of the alga *Chlamydomonas*,[30] and the Swedish team of Sylvan and Margit Nass had done the same for mitochondria.[31] Why would mitochondria and chloroplasts have their own DNA? Ris, Plaut, and the Nass duo noted similarities between the general appearance of DNA in free-living bacteria and the DNA visible in their cross-sections of mitochondria and chloroplasts. In cautious, reserved terms, these authors went as far as to suggest that endosymbiosis be revisited as a possible explanation for the origin of both organelles.

Margulis wasn't interested in being cautious. She had a grand vision of the evolution of complex life, and the origins of mitochondria and chloroplasts were pieces of the larger puzzle she had begun to assemble in support of it. She would settle for nothing less than a complete picture of when and how the eukaryotic cell had evolved from interacting prokaryotes. Her canvas was the ~3.8 billion year history of life on Earth. What was remarkable and truly unique was how she strove to glean and synthesize evidence from so many different realms of science: geology, palaeontology, ecology, bacteriology, genetics, and especially cell biology. As she did so, Margulis was heading towards what would become her first foray into publication hell—most researchers (and presumably the anonymous reviewers of her ill-fated 1967 paper) simply could not see her forest for the trees. To

Margulis, the forest was comprised of symbiotic microbial communities; the concept of the individual organism was of limited utility.

In a 2011 interview by Dick Teresi,[32] Margulis was asked if she ever got tired of being considered controversial. By this time her reputation as a dogmatic iconoclast was legendary. She had fought her whole career against what she called the 'neo-Darwinist population-genetics tradition', critical of the notion that mutations arising from within the cell could alone give rise to the astonishing diversity seen in the living world: 'natural selection eliminates and maybe maintains, but it doesn't create'. Only symbiogenesis, she argued, could do that. She pushed this idea to the limit in her 2002 book *Acquiring Genomes*,[33] published with her son Dorion. Margulis and Sagan proposed that symbiotic mergers between genetically distinct organisms underlay the process of speciation across *all* of life. It was a difficult pill for many researchers to swallow, in particular those focused on the biology of plants and animals.*

Equally difficult for most biologists to accept is Margulis's long-standing hypothesis that eukaryotic flagella—the whip-like appendages that nucleus-containing cells use to sense their environment and propel themselves through liquids—evolved from a symbiotic spirochete bacterium.† As their name suggests, spirochetes exhibit a corkscrew-like appearance and use a twisting motion to move about; they are also known to attach themselves to the surface of certain unicellular eukaryotes, where they indeed look much like flagella. But despite much searching DNA has never been found in association with the flagellar apparatus, and few researchers now believe, as Margulis steadfastly maintained, that once free-living spirochetes

---

* It is a rare book in which the author of the Forward—in this case Harvard University's Ernst Mayr (1904–2005)—seems to disagree with the main conclusions of the authors whose work she/he is 'endorsing'. *Acquiring Genomes* is one of them.
† Sperm tails are a classic example, as are the 'ciliary' appendages of the cells that line our respiratory tract. Note that Margulis abhorred the term 'flagellum' as applied to the motility apparatus of eukaryotic cells. She preferred to call this structure the 'undulipodium' so as to avoid comparisons with the bacterial flagellum, which is very different in form and protein composition from its eukaryotic counterpart.

played a role in the origin of eukaryotic motility. Fewer still are comfortable with her contentious proposal that the human immuno-deficiency virus does not cause AIDS, symbiotic spirochetes do.[34] In the words of Joanna Bybee, for Margulis 'no subject was too sacred to go untouched and unquestioned'.[35]

On the question of controversy, Margulis answered in a manner that came as no surprise to anyone who knew her: 'I don't consider my ideas controversial. I consider them right'. It was against the backdrop of her original formulation of eukaryotic cell evolution that the endosymbiont hypothesis was put to the test, and in the case of mitochondria and chloroplasts, Margulis would prove to be contro-versial *and* right. The results were spectacular. Hidden in the DNA of their remnant genomes, biologists would find the answer to the question of their evolutionary origins. The key to accessing this information was forged with the tools of molecular biology, the same tools that were used to usher in an exciting new era of scientific exploration: microbial phylogenetics.

4

MOLECULAR RULERS OF LIFE'S KINGDOMS

Imagine it's the year 2029. The United States is ruled by an iron-fisted authoritarian government, and as a 'humane' alternative to execution, political dissidents are sent to a special penal colony. What's special about the colony is that it exists in the past. Using a time machine invented by an eccentric mathematician named Edmond Hawksbill, rebels are banished to the Paleozoic Era, hundreds of millions of years ago. The 'criminals' have been accumulating there for decades; they are occasionally sent medical equipment and supplies for expanding their motley collection of huts, but are otherwise left to fend for themselves. Fortunately there are no large creatures around to prey upon them. Unfortunately, their diet consists mainly of brachiopod stew and trilobite hash. The so-called Cambrian explosion is still underway—the animals with which we humans are most familiar have yet to evolve; colonization of the land has only just begun. In terms of multicellular life, the rocky, depressingly barren surface of the Earth boasts little more than the odd patch of moss. With nothing to do but 'fish' and talk politics, the prisoners suffer in various stages of madness, longing for the past in a future to which they cannot return.

This bleak vision of a living death sentence is the premise of *Hawksbill Station*, a science fiction novel published in 1968 by the American writer Robert Silverberg.[1] For sci-fi fans with a penchant for time travel it's a good read. For geologists and evolutionary biologists, it's the ultimate thought experiment. Imagine being able to see what planet Earth was like half a billion years ago, what it was *actually like*. How many landmasses were there? What was the chemical composition of the atmosphere? And how well do our scientific predictions stack up against the actual data? Imagine if we could study real, live organisms from the ancient past. We could document their behaviour and ecological distributions; examine the structure of their cells; we could even study their genes. It would be infinitely more informative and exciting than bringing extinct organisms back to life *à la Jurassic Park*.*

As of this writing time machines do not exist—we are stuck in the here and now, and the best thing we've got to help us infer the history of life is the life that surrounds us. Of course, some of these organisms actually *are* from the past. We've learned a great deal about Earth's history from studying fossils, the remains of expired creatures 'trapped in time'. High up in the Canadian Rockies in British Columbia, for example, lies the Burgess Shale, a 500 million year old rock formation derived from what was once the ocean floor. It harbours a collection of fossilized life forms that at first glance could easily pass as science fiction: bizarre shrimp-like animals more than a metre in length; worm-like organisms with 'legs' and 'spines' but no obvious head; and of course trilobites, the segmented, hard-shelled creatures that were among the world's first arthropods (invertebrates with external skeletons and jointed appendages). The Cambrian Period, to which the Burgess Shale belongs, is of immense scientific importance: it corresponds to the 'brief' 60 million year stretch of time during which complex animal life burst on to the scene. Strange though they were, the Cambrian fauna appear to have been the source of most of the fundamental body plans exhibited by today's animals.[2]

---

* Nitpickers will of course question the science behind Silverberg's fiction. For example, was there enough oxygen in the Earth's atmosphere back then for humans to even breathe? And in case you're wondering, there are no women at Hawksbill Station. In order to avoid 'contaminating' the timeline, female political radicals are sent to a different colony, millions of years away from the men.

Yet fossils have their limits, and the older they are the less confidence we generally have in what they tell us. The chance that an organism will perish under just the right environmental conditions such that it will become a fossil is vanishingly small (as is the chance that it will ever end up in the hands of a human being). The odds are smaller still for soft-bodied creatures that lack mineralized, decay-resistant bones or shells. In the case of the Cambrian 'explosion', scientists have for many years debated whether it was a real event or the result of preservation bias—an artefact of the organisms that have, and have not, revealed themselves for study. Darwin himself struggled to reconcile the apparent 'sudden' emergence of animals during the Cambrian with the notion that the process of evolution leads to a gradual increase in organismal complexity over time.

Trying to make heads or tails of fossils taken from the so-called Ediacaran Period (635–542 million years ago), which immediately precedes the Cambrian (541–485 million years ago), has proven difficult. Globular clusters of cells apparently 'frozen' in the act of dividing have been found in the Doushantuo Formation in Southern China; based on the inferred patterns of cell division, they have been interpreted as being ~570 million year old animal embryos, a stunning discovery if ever there was one. But an alternative explanation is that they are not animals at all but in fact colonial algae.[3] Disc-shaped, striated Ediacaran fossils come in various sizes and *might* represent the squashed remains of marine invertebrate animals such as jellyfish. But can we be certain? They are simply too 'simple' to know for sure.

Palaeontologists generally know a fossilized cell when they see one.* And as remarkable as it seems, with advanced imaging techniques the signature nucleus of a fossil eukaryote can sometimes be observed. But in the absence of interpretable morphological features, the discovery of 'small round cells' in fossiliferous rock is of limited

---

* A curious exception was the 1996 announcement by NASA scientists of the discovery of putative bacteria in a Mars meteorite. Rod-shaped, yes, but these cell-like 'fossils' are much, much smaller than any known bacterium. Very few scientists are convinced of their authenticity.

scientific value, even if we know how old they are. In the case of microorganisms, 'small' and 'round' is often all we've got.

What if we didn't need to consider what an organism looks like in order to figure out where it came from? As touched upon in Chapter 2, we don't. The linear sequences of nucleotides in DNA and amino acids in proteins contain, in the words of Francis Crick, 'vast amounts of evolutionary information'. Extracting and interpreting that information is, in a sense, the next best thing to building a time machine—it opens a window on to the past history of living beings in a way that traditional morphology-based approaches cannot. Our conception of the tree of life was once based primarily on consideration of macroscopic life forms, namely animals, plants, and fungi. Through the use of molecules as evolutionary markers, the hidden microbial majority of our planet has taken its rightful place on that tree. It didn't happen overnight, but the result has been a transformation in our understanding of the very foundations of evolution.

The (RNA) catalogue of life

On the morning of Thursday 3 November 1977, the front page of *The New York Times* prominently displayed a picture of a scientist at work. It was not the sort of image most would conjure. For one thing, there's no lab coat. It shows a middle-aged man in jeans and sneakers sitting in his office—his feet are up on the desk, his hands are gesticulating, his face purposeful. In the background is a chalkboard festooned with notes and diagrams. The title of the article—Scientists Discover a Form of Life That Predates Higher Organisms—was provocative yet cryptic, like the discovery itself. And the picture speaks volumes about the life and times of University of Illinois microbiologist Carl Woese (1928–2012), champion of molecular evolution and discoverer of the third kingdom of life.

That chalkboard was everything to Woese—brainstorming was part of his day-to-day existence. Habitually drawn to big problems, he needed to push both experimental and theoretical frontiers in order to get where he wanted to go. Woese's scientific passions revolved around translation, the process by which cells make the proteins needed to sustain life.

Recall Francis Crick's 'central dogma' of molecular biology: DNA makes RNA makes protein. At the heart of protein synthesis lies the genetic code, the complete set of three-nucleotide codons that serves as the bridge between an organism's genotype and its phenotype.* A gene is a linear stretch of DNA that 'codes' for a specific protein; in order to produce that protein, the cell must first synthesize (or 'transcribe') a protein-specific messenger RNA (mRNA) molecule. It does this by copying the sequence using the DNA as a template. It is the resulting mRNA, and the amino acid-specific codons it contains, that serves as a molecular 'ticker tape' for the ribosome, the protein synthesis factory of the cell. Working its way along the mRNA, the ribosome reads the nucleotide triplets and strings together the corresponding linear chain of amino acids—the base units of protein. Once released from the ribosome, the newly built protein folds into its biologically active 3D structure and is free to fulfil its cellular role.

Ribosomes are massive, abundant, extraordinarily conserved macromolecular complexes; their make-up is very similar from organism to organism. The 'ribo' part of the name comes from the fact that, in addition to containing dozens of proteins, ribosomes possess several different types of ribonucleic acid (RNA) molecules.[4] These so-called ribosomal RNAs (rRNAs) are very different from mRNAs: they serve not as transient carriers of genetic information but as functional elements whose primary sequence and 3D shape are essential to their job. What Woese did was to crack open the ribosome and reveal the essence of its RNA core. In doing so, he realized the potential of rRNA to serve as a universal molecular chronometer for the study of evolution.

---

* An organism's 'genotype' is its genetic makeup, its complete heredity information. 'Phenotype' refers to an organism's physical properties.

Woese once described himself as 'a molecular biologist in search of Biology'.[5] He was trained to think 'nuts and bolts', having studied physics as an undergraduate at Amherst College and, in the early 1950s, biophysics as a PhD student at Yale University. But as his career progressed Woese felt increasingly strongly that the discipline of molecular biology had become too reductionist, too myopic. With its high-throughput experimentation and advanced biotechnological capabilities, modern biology was, he felt, losing sight of its ultimate purpose. Biology—with a capital B—needed to go back to basics, back to Darwin: 'You can't understand the gene without understanding translation, and you can't understand translation without understanding its evolution'.[6] This involved taking into account all of life, not just the organisms we can see.

In the 1960s, with the burgeoning field of molecular biology still basking in the glow of the double helix, Woese had set out to probe the fine-details of protein synthesis from the broadest possible perspective. How is the translation apparatus able to do what it does? How did the universal genetic code evolve? Was it simply a 'frozen accident', as Crick had postulated, or does its precise make-up, together with the inner workings of the ribosome, hold the key to understanding the earliest events in the evolution of the cellular world? Tackling these problems led Woese into the realm of microbiology, to investigate the nature of the bacterium and its place in nature.

At the time the bacterium didn't really *have* a place in nature, or at least not one that scientists generally agreed on. *Bergey's Manual*, the long-standing Bible of microbiology named after the American David Hendricks Bergey (1860–1937), was first published in 1923 as a guide for 'determinative bacteriology'—in essence a comprehensive resource for the identification of unknown bacteria. It served (and indeed still serves) as an invaluable tool for those working on the front lines. Physicians, for example, needed to know what they were up against; they needed to be able to identify disease-causing microbes and place them within a framework of existing knowledge. Of course, in order to be able to communicate and share information about microbes, scientists needed to call them something; the microbes needed to be given labels. On what criteria those labels should be based was an open question.[7]

Within the domains of botany and zoology, the so-called Linnaean system of taxonomy has long served as *the* framework for organismal classification. It is a rank-based scheme stemming from the works of eighteenth-century Swedish botanist Carl Linnaeus (1707–1778): each and every organism is given a binomial genus-species designation, which is itself nested within a larger framework of family, order, class, and so on. We humans, for example, belong to the species *sapiens* within the genus *Homo*. Together with our gorilla, chimp, and orang-utan relatives, we make up the family Hominidae.

The classical Linnaean system has served science well for over 150 years. But as logical as it is, it can be difficult to apply to organisms that aren't easy to see, things that don't have wings, legs, or leaves. Beginning in the 1940s, there was increasing debate over how to properly classify microbes, and indeed whether they *could* be classified in the same way that plants and animals were. Should such a classification scheme attempt to be explicitly 'natural'—i.e., convey information about evolutionary relationships? Does it make sense to talk of a bacterial 'genus' and 'species'? Could a natural taxonomy of microorganisms, even if it exists, actually be uncovered? An approach to taxonomy based on the features of molecules had the potential to address these thorny issues.*

Woese sought to develop a system for the natural classification of bacteria based on the evolutionary information contained within rRNA. As we have learned, Zuckerkandl and Pauling had in the early

---

* It is worth noting that in the era of comparative genomics, biologists still debate the existence of microbial species, albeit for different reasons. As discussed in Chapter 2, microbes readily exchange genes across species boundaries, which serves to 'scramble' the evolutionary history of the set of genes harboured by a given organism. The extent to which this renders the concept of a bacterial species meaningless depends in part on how one defines a 'species'.

1960s proposed that protein sequences change over time in a clock-like manner and that this could be a useful feature for taxonomy: the lower the degree of amino acid sequence similarity shared between the proteins of different species, the greater the amount of time that had passed since they diverged from a common ancestor. In the case of Woese's rRNA, it was the nucleotide sequences whose divergence levels could be assessed. In a letter to Francis Crick in 1969, Woese referred to the use of molecular evolutionary reconstruction as accessing the cell's 'internal fossil record'.[8] Not only would these rRNA 'fossils' be information-rich, but finding them was a sure thing. Ribosome-based protein synthesis lies at the very core of cellular life; every organism on Earth, no matter what it is, where it lives, or how it makes a living, has rRNA genes that could in principle be examined.

Ribosomal RNA possessed all the features one could hope for in a universal molecular tracker of evolution, but there was a problem: a method for determining the primary sequence of a DNA or RNA molecule did not exist. Thanks to the work of Cambridge biochemist Frederick Sanger in the 1950s, the amino acid sequence of a protein could be obtained, and by the mid-1960s, Sanger's team had figured out how to characterize short snippets of RNA from model species such as yeast and E. coli.[9] Building on these methods, Woese devised a set of procedures for sequencing rRNA that became known as RNA cataloguing, or 'oligonucleotide cataloguing'. (The term oligonucleotide refers to a short fragment of DNA or RNA, typically less than a dozen nucleotides long.) In its original form it was about as 'powerful' as one of van Leeuwenhook's early microscopes, but it provided an all-important first glimpse of the genetic diversity contained within the microbial biosphere. All told, there were about half a dozen people on the planet who knew how to do it.

Woese's procedure took advantage of the extraordinary abundance of ribosomes—and thus rRNA molecules—in active cells. Like Sanger, Woese and his team began by growing bacterial cultures in the presence of radioactively labelled phosphate. The idea was that as cells continuously built fresh ribosomes to replace old ones, the radioactive compound would be incorporated into the nucleotide precursors of RNA and then into newly synthesized rRNA. Once the bacterial cultures became suitably dense, total cellular RNA was isolated and fractionated using a technique called 'gel electrophoresis', whereby a biological sample is embedded within a porous substance the consistency of a gelatin-based dessert. When an electric field is applied, the gel acts as a sieve, separating the complex mixture of molecules contained within it based on their relative size and charge. Because of their ultra-high abundance, rRNAs could be detected and separated from the other types of radioactive RNA present in the sample, such as mRNA. Purified rRNA in a test tube—so far, so good.

The next step was the key to obtaining actual sequence information. The purified rRNA molecules were broken into shorter fragments by exposing the sample to specific RNA-degrading enzymes, such as those that cut RNA strands after the ribonucleotide G. The process of electrophoresis was then repeated, this time with the chewed-up rRNAs loaded on the gel. The prediction was that rRNA molecules from different bacterial species should differ in primary sequence, and consequently, yield different-sized pieces and gel migration patterns. The more distantly related the species, the greater those differences should be. This proved to be the case. Woese learned to make sense of these variations, which manifested themselves as differences in the patterns of spots on X-ray film. The patterns were as consistent and predictable as the species-specific decorations on the wings of a butterfly.

At first Woese's catalogue of life was comprised of rRNA profiles for only a handful of bacteria. The methods were crude, painstaking, expensive, and by today's standards dangerous (they required levels of radioactivity orders of magnitude above what would be approved for use in a modern laboratory). But the results were promising, to Woese at least. As the procedures were fine-tuned, Woese and his colleagues became increasingly fluent in the language of RNA. The differences between eukaryotic and prokaryotic rRNA profiles could be spotted at a glance, and a rough 'tree' representing genetic relationships between different bacteria could be produced by comparing and contrasting their rRNA oligonucleotide sets.[10]

What quickly became clear was that Woese's rRNA-based tree was at odds with the classification scheme laid out in *Bergey's Manual*, which in the 1970s clustered bacteria into groups based mainly on what little morphological differences could be gleaned from them: whether they were spiral, round, or rod-shaped, whether or not they had flagella protruding from their surface, whether they were capable of going dormant by forming spores, and so on. The reason(s) for the discrepancies between the two approaches to classification—molecules versus morphology—was at the time far from clear. But Woese's data served to rekindle the debate over whether a natural phylogeny of microbes was a realistic goal. Woese himself was undeterred. The more RNA samples from diverse bacteria that made their way through his lab's pipeline, the more he believed he was on to something big, something that had the potential to transform biology.

Kingdoms and molecules

In 1976 Woese analysed the RNA sample that would change his life— and land him on the cover of *The New York Times* the following year. The source was an enigmatic bacterium by the name of *Methanobacterium thermoautotrophicum*. This particular microbe had been made available to Woese by his University of Illinois colleague Ralph Wolfe, who, as a biochemist-microbiologist, was trying to understand how this methane-producing species managed to thrive in sewage sludge without oxygen and at a temperature of 65°C. Wolfe figured that if any bacterium could benefit from a bit of evolutionary enlightenment it was this one. Given its natural habitat, it is perhaps not surprising that *Methanobacterium* proved difficult to tame in the lab. But once Wolfe's group managed to optimize its growth conditions, RNA was extracted and processed by Woese's team. As always, the X-ray films ended up in the hands of Woese himself for analysis and interpretation.

He could tell immediately that something was different. Where were the signature spots diagnostic of bacterial rRNA? Some were there but others were not, and eukaryote-specific signatures could also be seen. This was, of course, impossible. Assuming that something had gone wrong experimentally, Woese ordered the experiments repeated from scratch. The results were the same. When similar rRNA profiles were retrieved from other methanogenic species, he ceased with the scepticism and excitedly began working through various scenarios that could explain the data in hand. What on earth had they discovered?

In essence, Woese and colleagues had discovered that the methanogens were a group of organisms that *looked like* bacteria but were *not* bacteria. They showed that there were in fact two distinct types of prokaryotic cells in the world: 'eubacteria', the 'true bacteria' that have entered into our discussions thus far, and so-called 'archaebacteria'. It must be said that Woese himself would have objected to my describing it this way, for he believed that the archaebacteria—or archaea as they were subsequently re-named*—were as distinct from 'typical' bacteria as they were from eukaryotes. Together with his postdoctoral fellow George Fox, Woese proposed that at the deepest level there were not two but three 'kingdoms' (or domains) of life: all living beings fell into one of three 'aboriginal lines of decent', of which the archaea were considered to be the most ancient.[11] To Woese, lumping bacteria and archaea together as prokaryotes was unnatural. The term 'prokaryote' was considered to be devoid of evolutionary meaning; he and his followers would eventually suggest that it be removed from scientific discourse.

These were controversial claims to say the least. Certainly no one could deny that with respect to habitat and lifestyle, the archaea brought new meaning to the term 'diversity'. In addition to methane-belching

---

* The term archaea comes from the Ancient Greek meaning 'ancient things'. Both archaebacteria and archaea are used in the scientific literature; the latter is now more common.

anaerobes like *Methanobacterium*, the archaea were found to encompass all manner of 'extremophiles'—these include salt-loving members of the genus *Halobacterium* (which happily live in the Dead Sea); acid lovers such as *Thermoplasma* and its kin; and hyperthermophilic organisms such as *Pyrodictium*, which live in deep-sea vents at temperatures over 100°C.*

In many respects Woese was spot-on about the distinctiveness of the archaea. The chemical composition of their membranes, for example, is unique in the natural world. Comparative genomic investigations have also revealed that archaea share many molecular features with eukaryotes to the exclusion of bacteria, including certain components of their DNA replication, transcription, and translation machineries. In other ways, however, archaea and bacteria are biochemically much more similar to one another than either of them are to eukaryotes. And an important fact remains: at the level of cellular organization archaea are prokaryotes, plain and simple. Distinct though they may be, many researchers today still consider archaea to be 'just bacteria', as the vast majority of scientists did when their existence was first announced to the world in 1977.

Reclusive by nature, Woese attended scientific conferences only rarely. But while he seldom defended his ideas in public, he never shied away from the chance to argue in print for the importance of microbes in general and archaea in particular. In 1998 his views on the subject were passionately laid bare in the pages of *The Proceedings of the National Academy of Sciences (USA)*, in what would be the final round of a long-standing debate with famed Harvard evolutionary biologist Ernst Mayr (1904–2005).† An ornithologist by trade, Mayr had for many years been the most prominent (if not the most qualified) critic of Woese's three-domains view of life. Mayr likened Woese's discovery

---

* Deep in the ocean it is possible for organisms to survive at temperatures greater than 100°C because the increase in pressure means that the boiling point of water is increased.

† Prodigious doesn't even begin to describe Mayr: he published more than 25 books and 700 scientific articles up until his death at the age of 101.

of the archaea to the 'discovery of a new continent'—high praise indeed—yet he simply could not fathom why Woese and company chose to weigh the bacterial-archaeal divergence so heavily: 'The evidence...shows clearly that the archaebacteria are so much more similar to the eubacteria than to the eukaryotes that their removal from the prokaryotes is not justified'.[12] Yet it was not, according to Woese, simply a 'taxonomic quibble...the disagreement between Dr Mayr and myself is not actually about classification. It concerns the nature of Biology itself. Dr Mayr's biology reflects the last billion years of evolution; mine, the first three billion'.[13]

5

BACTERIA BECOME ORGANELLES

An Insider's Take

The 'Woesian revolution' had a profound impact on the nature of biological research. It was in one sense a methodological revolution: the decade-long commitment of Carl Woese to acquiring ribosomal RNA (rRNA) sequence data from as many prokaryotes as he could get his hands on yielded the very first all-embracing, genotype-based classification scheme for microorganisms.* It served as an important proof of concept whose legacy is incontrovertible: rRNA is still the go-to gene for molecular systematics. It is typically the first molecule scientists turn to when they need a quick and definitive answer to the question 'what are you?' for a newly discovered organism—not just for microbes, but for *any living thing*, big or small, prokaryote or eukaryote. And, thanks to the pioneering efforts of Norman Pace of the University of Colorado, rRNA is increasingly used to answer the question 'who's there?' of any environment one wishes to explore, natural and otherwise: the ocean, a farmer's field, polar ice sheets, computer keyboards and shower curtains, every nook and cranny of the human body. The discipline of microbial ecology, in which biologists address fundamental questions about the structure

---

* Woese's molecular phylogenies were in fact among the first to be done for *any* group of organisms. As discussed in Chapter 2, in the 1960s Zuckerkandl, Pauling, Fitch, and other researchers used protein sequences to infer evolutionary relationships among animals.

and dynamics of microbial communities, was founded on the use of rRNA sequence analysis. Simply put, small-subunit rRNA is the most widely sampled gene on the planet and probably always will be.

Woese's revolution was also conceptual. As we have learned, his gene-centric approach to microbial systematics revealed a deep divergence within the prokaryotes—so deep that at its foundation life is now generally conceived as comprising three domains: bacteria, archaea, and eukaryotes. Regardless of what archaea are or are not, taxonomically speaking, it is clear that there is far more genetic and biochemical diversity within the microbial biosphere than in all the macroscopic organisms on Earth. The significance one chooses to place on this hidden diversity, relative to the morphological variation so readily apparent in the plant and animal world, is largely a matter of perspective. But we can no longer ignore what Woese referred to as 'biology's sleeping giant';[1] any account of the patterns and processes of evolution that fails to properly consider the microbial realm must be considered incomplete.

Woese would have dropped his rRNA cataloguing technique 'like a hot potato' if he could. Linda Bonen would know—she was his research assistant in the early 1970s.[2] Well before the discovery of the archaea, Bonen, now a Biology Professor at the University of Ottawa, had toiled in Woese's lab at the University of Illinois, testing and troubleshooting, refining the protocols so as to maximize the amount of useful information that could be gleaned from his beloved X-ray films. Woese, of course, didn't quit; the sequence data he craved was far too important and exciting to wait around for. As inevitably happens in any technology-driven endeavor, rRNA cataloguing was eventually supplanted by easier, faster, and more accurate sequencing methods. But in the mid-1970s it was the only game in town, and big problems in biology were ripe for the picking.

Foremost among them was the origin of the complex cell. If, as Woese and colleagues had shown, the evolutionary history of cellular lineages could be revealed using rRNA-based phylogeny, then perhaps the origins of the mitochondria and chloroplasts residing within the eukaryotic cell could be as well. Bonen took the rRNA cataloguing technology she had learned from Woese north of the border to Dalhousie University in Halifax, Nova Scotia, Canada, where two research groups began pioneering experiments on the nature of orga- nellar rRNA. The year was 1974—four years after Lynn Margulis had brought endosymbiosis to the scientific mainstream with the publi- cation of her now famous book, *Origin of Eukaryotic Cells*.[3]

Low hanging fruit on the tree of life

It was Woese who suggested to Bonen that she work with Ford Doolittle, who was a new recruit to Dalhousie.[4] Doolittle had grown up in Urbana, Illinois, the son of an artist-professor and best friend of one Will Spiegelman. Will's father happened to be a biochemist. Among other scientific contributions, Sol Spiegelman (1914–1983) developed the technique of nucleic acid 'hybridization', whereby sin- gle-stranded DNA or RNA molecules could be made to form double helices in the test tube by the Watson–Crick-style base complemen- tarity discussed in Chapter 2. The extent to which a DNA/RNA mol- ecule 'zips' itself up in this fashion can be measured experimentally, and hybridization is now a cornerstone of modern molecular biology.

Doolittle's first exposure to the exciting world of research was as a high school student, 'washing dishes and growing *E. coli*' in the Spiegel- man lab during summer vacation.[5] Here he was inspired by the bright young scientists who flocked to Urbana from around the globe. The experience was transformative: 'They shared with me the triumph of the elegant hypothesis confirmed, the dismay of the experiment scooped, and most of all the complex joy of figuring out how living things work'.[6]

Having left Urbana to pursue undergraduate and graduate training at Harvard and Stanford, respectively, Doolittle returned to Spiegel- man's group at the University of Illinois as a postdoctoral fellow in 1968. It was here that he first met Woese, whose lab was just down the hall and who by that time had begun slogging away at rRNA cataloguing. Although their scientific interests overlapped, Doolittle and Woese did not collaborate—mostly they drank beer together.[7] From Urbana, Doolittle went to the Denver Colorado laboratory of Norman Pace, himself a former trainee of Woese, where he studied the synthesis of rRNA in the bacterium E. coli. Doolittle started his own lab in Dalhousie's Department of Biochemistry in 1971 with the mind to delve into the mechanics of rRNA synthesis in much greater detail.[8]

Doolittle was not an evolutionary biologist, not by any stretch of the imagination. But he had read *Origin of Eukaryotic Cells* and a colleague in his department, Ian McLean, happened to be growing cyanobacteria for studies on the biochemistry of photosynthesis. When Bonen joined Doolittle's lab in 1974, it dawned on him that the proposed evolutionary connection between cyanobacteria and the chloroplasts of plants and algae that had been so passionately spelled out by Margulis could be tested using their collective laboratory skills.[9]

Mitochondria were brought into the Dalhousie mix by the Canadian Michael Gray, who arrived in the Department of Biochemistry in 1970, the year before Doolittle. Gray's name is now synonymous with mitochondrial evolution, but it certainly didn't start out that way. He was a nucleic acid biochemist. During postdoctoral research at the Stanford University School of Medicine, Gray had studied DNA repair systems in E. coli, an exciting but fiercely competitive area of research at the time. If he were going to be successful he would need to carve out a niche for himself. That niche turned out to be mitochondrial nucleic acids.

From his graduate student days at the University of Alberta, Gray recalls spending many an evening in the lab babysitting a floor-to-ceiling contraption called a sephadex column—basically a giant sieving device used for the isolation of transfer RNAs (tRNAs), the small molecules whose cellular job is to bring amino acids to the ribosome for protein synthesis. Late one night in 1967 Gray was passing the time browsing the latest issue of *The Proceedings of the National Academy of Sciences (USA)*. He stumbled upon an article entitled 'Mitochondrial Transfer Ribonucleic Acids' by Edgar Barnett and David Brown,[10] which described the characterization of tRNAs from the mitochondria of the fungus *Neurosposa*. The paper struck a chord: 'why on earth would mitochondria need a translation system?'[11]

Gray knew what mitochondria were. But like most biochemists in the late 1960s he was oblivious to the long-standing debate over their evolutionary origin—he had never heard of the endosymbiont hypothesis—and had no idea why mitochondria would possess the ability to make their own proteins. Based on their 1967 paper, Barnett and Brown didn't seem to either. When the time came to establish his own group, Gray saw an opportunity and took it: 'I determined that virtually nothing was known about plant mitochondrial rRNA and tRNA and decided to make them the focus of my research program'.[12] He had never actually worked with mitochondria, but he had an endless supply of starting material in the form of giant sacks of wheat germ obtained free of charge from a local mill. All he had to do was learn how to purify them from whole cells. It was the beginning of a lifelong fascination with the 'powerhouse' of the eukaryotic cell.

Were it not for Bonen joining his lab, Doolittle reckons he would never have begun working on problems in evolutionary cell biology, the area that would ultimately define his career.[13] Gray too probably would not have done so without the efforts and enthusiasm of his first PhD student, Scott Cunningham. The work of these four scientists on mitochondrial and chloroplast evolution initially had little to do with their mainstream research—it was a side project, nothing more. To obtain funding specifically to delve into the evolutionary history of these organelles would have been exceedingly difficult. The topic was too controversial, too 'fringe', certainly not something that a card-carrying biochemist would generally admit to being interested in.

How to test the endosymbiont hypothesis

While Doolittle and Gray came at the problem of organelle evolution from the bottom up, the approach of F.J.R. (Max) Taylor, Professor Emeritus of Biological Oceanography and Marine Phytoplankton at the University of British Columbia, was decidedly top down. Taylor's scientific interests were not molecular or biochemical. He was focused squarely on cells as organisms: their structure, ecology, and evolutionary history. Born in Cairo, Egypt, and raised in South Africa, Taylor has a PhD from the University of Cape Town; he has for many years been the world's leading authority on a bizarre group of unicellular 'algae' called dinoflagellates. I say 'algae' because only some dinoflagellates photosynthesize for a living. Those that do are notorious for causing 'red tides', giant algal blooms in which the cells grow to such high concentrations that they colour the water and become toxic to marine animals and the humans that eat them.* Taylor has studied this remarkable phenomenon at research stations the world over.

Less well known is the fact that dinoflagellates are among the most fascinating examples of 'evolution in action' to be found in all of biology. For many millions of years dinoflagellates have made it their business to ingest other organisms. In and of itself such behaviour is not unusual; many unicellular eukaryotes obtain their nutrition in this manner. But dinoflagellates seem to have trouble digesting their food—they have acquired photosynthesis over and over and over again throughout their evolutionary history by commandeering the light-harvesting capabilities of their algal prey. Coupled with his curiosity of all things eukaryotic and microbial, it was through his studies of algae and their curious habit of housing foreign cells within their cytoplasm that Taylor came to be interested in the question of organelle evolution.

Margulis receives the lion's share of the credit for having thrust endosymbiosis back into the scientific foreground of the late 1960s. But she was far from alone in her efforts to account for the internal complexity of the eukaryotic cell. Other researchers had begun

---

* Paralytic shellfish poisoning can occur by ingestion of filter-feeding bivalve molluscs (e.g. clams and mussels) in whose tissues an algal-produced toxin has accumulated. Algal blooms can become so vast that they are visible from outer space. Despite their name, red tides are only occasionally red and they have nothing specifically to do with the tide.

publishing on the topic at the same time and in high-profile venues to boot. In 1967, the same year as Margulis's (Sagan's) seminal contribution, the Norwegian microbiologist Jostein Gøksoyr (1922–2000) penned a brief scenario in the pages of *Nature* depicting how eukaryotes might have evolved from symbiotic associations between multiple prokaryotic lineages.[14] In 1970 the celebrated American botanist Peter Raven published a thorough account of the origins of mitochondria and chloroplasts in the prestigious American journal *Science*, concluding that: 'Many independent symbiotic events may have been involved in the origin of these cellular organelles'.[15] Yet these authors received (and still receive) little in the way of recognition for these specific contributions; it was primarily around Margulis and her broad-sweeping hypothesis for early eukaryotic evolution that intense scientific debate began to swirl. At stake was arguably the single most important piece of the puzzle of how complex life on Earth had evolved.

Those in favour of endosymbiosis endorsed what was referred to as the 'xenogenous' (or foreign) origins scenario: mitochondria and chloroplasts were derived from once free-living prokaryotes; they had come from outside the eukaryotic cell. Vigorous counterarguments were made by various researchers, most notably Rudolph Raff, Henry Mahler, and Allan Allsopp. These authors posited that mitochondria and chloroplasts were of 'autogenous' origin—they had evolved from *within* the confines of the cell (Fig. 3).[16]

Taylor was sympathetic to both viewpoints. He'd spent years studying natural communities of algae, countless hours peering down the microscope; he knew how common it was for microorganisms to form intimate associations with one another in nature. For Taylor, endosymbiosis wasn't an abstraction—he had seen it with his own eyes. Yet he was also inherently cautious and speculation-averse, a mindset that had been reinforced during his early research training. It was his nature to want to explore a scientific problem from all possible angles before making up his mind.[17]

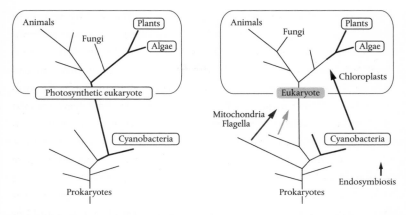

Fig. 3. Schematic evolutionary trees depicting the classical ('autogenous'; left) and symbiotic ('xenogenous'; right) views of the evolution of eukaryotic organisms from prokaryotic cells. Diagrams are based on those presented by Lynn Margulis in *Origin of Eukaryotic Cells* (1970). The autogenous model proposed that the ancestral eukaryotic cell was photosynthetic, having evolved from a cyanobacterium-like prokaryote. Under the autogenous model, the internal features of the eukaryotic cell evolved from within the cell, and were inherited in a vertical fashion. In contrast, the xenogenous model of eukaryotic evolution proposed by Margulis and others postulated that mitochondria and chloroplasts were derived from prokaryotes by endosymbiosis. Margulis also believed that the flagella of eukaryotes were of endosymbiotic origin. In both diagrams the photosynthetic lineages are highlighted. Arrowheads show putative endosymbiotic events.

And so as the debate over the origins of mitochondria and chloroplasts intensified, Taylor became something of a moderator-referee, serving as both a critic and confidant of Margulis. Interestingly, he had never heard of her until 1973—some years after her now famous early publications—when she invited him to speak at a conference in Boulder, Colorado. It was here that Taylor first learned of the similarities between his own ideas about eukaryotic cell evolution and those of Margulis and her contemporaries.[18] In his 1974 article stemming from the Boulder conference, Taylor presented what he referred to as the 'Serial Endosymbiosis Theory' (SET).[19] It was a landmark contribution. With clear, elegant prose, Taylor summarized and synthesized the evidence for and against xenogeny and autogeny (terms he himself introduced) in a manner that was much more 'accessible' to the average biologist than did the dense, jargon-rich works of Margulis. In so doing, Taylor highlighted those aspects of the endosymbiont hypothesis that were most amenable to rigorous scientific testing.

The facts presented in support of SET were essentially these: chloroplasts are structurally similar to cyanobacteria; algae and plants (and specifically their chloroplasts) carried out photosynthesis in a manner that is remarkably similar to cyanobacteria; mitochondria share structural and biochemical features with aerobic, free-living bacteria; both mitochondria and chloroplasts possess DNA that, when viewed under the electron microscope, is structurally similar to bacterial DNA; numerous examples of endosymbiotic relationships between eukaryotes and prokaryotes can be found in nature, many involving cyanobacteria.[20]

Did these facts prove that chloroplasts and mitochondria had evolved by endosymbiosis? No they did not. They were *consistent with* SET, but they could not prove it. Among other things, there were still many uncertainties locked within that 'black box' of cellular diversity, the prokaryote–eukaryote divide; the renowned microbiologist Roger Stanier (1916–1982), who was also an early endorser of endosymbiosis,[21] had referred to it as 'the greatest single evolutionary discontinuity to be found in the present-day world'.[22] Quite how cells had managed to get from one side to the other during the course of evolution was anyone's guess.

Regardless of which scenario was preferred, autogenous or xenogenous, no one with a stake in the controversy doubted that the eukaryotic cell had evolved from *some sort* of simpler prokaryote. It was a question of how, from whom, and what—if anything—photosynthesis had to do with it. The cyanobacteria (which at the time were misleadingly called 'blue-green algae') featured prominently in both models, but in very different ways. The autogenous model invoked strict vertical (non-endosymbiotic) inheritance of oxygen-producing photosynthesis from prokaryotes to eukaryotes (Fig. 3).

The Americans Richard Klein (1923–1997) and Arthur Cronquist (1919–1992) of the New York Botanical Garden were among its staunchest supporters; they had labelled the endosymbiont hypothesis 'a bad penny', one that had been in circulation for a long time.[23] In 1970 Klein summarized their preferred autogenous origins model thus: 'The conservative point of view is to assume that the main stream [of evolution] ran from the photobacteria to the unicellular blue-green algae [cyanobacteria] and thence to the eukaryotic unicellular algae'.[24] It was, Klein and Cronquist believed, most economical to postulate that all the internal complexity of the eukaryotic cell—the nucleus, the membrane-bound organelles, the rigid cytoskeleton, and so on—had evolved from simpler precursors contained *within* a cyanobacterium-like prokaryote.

Never one to mince words, Margulis referred to the notion of a photosynthetic eukaryotic ancestor as the 'botanical myth'.[25] If evolution had proceeded along a path leading directly from simple cyanobacteria to complex photosynthetic eukaryotes, then where were all the chloroplasts? Sure, plants and algae had them, but what about animals, fungi, single-celled amoebae, and so on? The autogenous model demanded that all present-day non-photosynthetic eukaryotes had lost their light-harvesting organelles somewhere along the line and evolved a new primary mode of energy acquisition (Fig. 3). Why such a large fraction of the known diversity of eukaryotic life would have abandoned a lifestyle as lucrative as photosynthesis was not clear.

For Margulis and her fellow endosymbiont enthusiasts, it was far 'easier' to assume that eukaryotic photosynthesis evolved well *after* the origin of the eukaryotic cell itself—it had been acquired on a specific sub-branch of the eukaryotic tree by a eukaryote ingesting and assimilating a cyanobacterium. SET readily accounted for the striking similarities between chloroplasts and cyanobacteria, as well as the 'patchy' distribution of photosynthesis across the tree of eukaryotes.

It is difficult to get biologists to agree on what is, and is not, 'easy' in evolution. Whereas the autogenous model for the origins of mitochondria and chloroplasts emphasized gradual change, endo-symbiosis was all about taking evolutionary 'leaps'. Such leaps tend to make biologists uncomfortable; which scenario was deemed most likely was greatly influenced by one's preconceived notions about the fundamental processes underlying evolution. Needless to say, the initial rounds of debate led to stalemate. No amount of arguing was going to change anyone's mind.

But there was one thing that scientists on both sides of the table could agree on: studies of molecular sequences had the potential to change everything. The eukaryotic cell contains two, and in the case of plants and algae three, DNA-containing compartments whose evolutionary histories could be inferred using the tools of molecular phylogenetics.* The autogenous and xenogenous models for organelle origins made very different predictions about how the genes residing in the nucleus, the mitochondrion, and (when present) the chloroplast should be related to one another. The endosymbiont hypothesis would stand or fall based on what these relationships proved to be.

Shining a light on chloroplasts

More data! We need more data! It's the battle cry of scientists every-where. When it comes to molecular phylogenetics, the more data you have the more accurate your evolutionary inferences will be. It will come as no surprise that the first full-length rRNA sequence was determined for the organism about which scientists know the most: the lab workhorse bacterium *E. coli*. What is remarkable is that it took researchers in the laboratory of Frenchman Jean-Pierre Ebel *ten years* to

---

* As we shall learn in Chapter 7, some eukaryotes contain four and even five genomes as a result of the 'capture-recapture' method of chloroplast evolution exhibited by dinoflagellates and other algae. Molecular phylogeny has been spec-tacularly successful in revealing 'who has eaten whom' in these multiple rounds of endosymbiosis.

complete the task. Hard-fought snippets of sequence had been available for some time, but it wasn't until 1978 that Ebel's team published the complete small-subunit (or '16S') rRNA molecule in the pages of the European journal *FEBS Letters*.[26] It was a significant landmark, and no one stood to benefit from it more than Woese. He had just announced the existence of the archaea, and was desperate for more data with which to shore up his results and silence his critics. The problem was he couldn't access the information; the October issue of the journal had yet to make its way to the University of Illinois. Woese couldn't wait. He picked up the telephone and called Doolittle at Dalhousie University. Doolittle read out the *E. coli* sequence to Woese over the phone—all 1,542 nucleotide letters of it.[27]

As Woese continued his never-ending quest for prokaryotic rRNAs, Doolittle and Bonen had begun charting new territory. They set out to determine whether the DNA that resides within algal chloroplasts was more similar to cyanobacterial DNA or to algal nuclear DNA. They were testing the idea that chloroplasts *used to be* cyanobacteria and were not simply cyanobacterial lookalikes. They were carrying out what they hoped would be the definitive test of the endosymbiont hypothesis.

The cyanobacterial sequences proved 'easy' to obtain. Doolittle grew cultures of cyanobacteria in the presence of radioactivity and purified the labelled rRNA, as he had learned to do in Colorado. He then handed them over to Bonen who, with encouragement and advice from her former mentor Woese, cut the molecules into smaller pieces, spread them apart in an electric current, and, together with Doolittle, puzzled through the constellation of X-ray spots. Short stretches of nucleotide sequences were painstakingly inferred.

Accessing and interpreting the rRNA sequences from the chloroplast was more complicated, residing as it did within a compartmentalized eukaryotic cell. Largely thanks to the work of Hans Ris and Walter Plaut in the early 1960s, it was known that chloroplasts contained DNA.[28] Nobody knew quite what to expect from the sequence of that DNA, but chloroplasts were also known to contain ribosomes—they could be seen with an electron microscope—and if ribosomes were present, then so too were ribosomal RNAs.

Bonen and Doolittle focused on what at first glance seemed like an unusual choice of organism, an obscure 'red' alga called *Porphyridium*. There was a method to their madness: *Porphyridium* was a unicellular species that could easily be grown to high density in the laboratory, which made it possible to obtain sufficient starting material for molecular experimentation. Red algae also figured prominently in autogenous scenarios of organelle evolution. By virtue of their relative morphological 'simplicity'* and certain features of their photosynthetic apparatus, Klein and Cronquist had postulated that red algae were a 'primitive' lineage of eukaryotes—more specifically, *the* lineage that linked the eukaryotic domain to cyanobacteria.[29] It was a proposal that could be tested using molecular data.

Bonen and Doolittle successfully characterized two evolutionarily distinct rRNA sequences from *Porphyridium* cells, the chloroplast and nucleus-derived versions of the molecule. The data were published in 1975 in *The Proceedings of the National Academy of Sciences (USA)*.[30] Fittingly, in the very same issue Woese and colleagues presented the chloroplast rRNA sequence of another alga, *Euglena*.[31] Together with an expanded set of cyanobacterial rRNAs published by Bonen and Doolittle,[32] these results spoke emphatically to the prokaryotic nature of chloroplasts: their rRNA sequences were more than 80 per cent *identical* to those of cyanobacteria. In stark contrast, the degree of similarity observed between the chloroplast and nuclear rRNA sequences was so low as to be difficult to discern with the fragmentary data available at the time.

---

* Red algae lack the flagella, or motility appendages, present in the vast majority of other eukaryotes.

The origin of mitochondria

Mitochondria represented the final piece of the puzzle. But it wasn't simply a matter of plugging them into place, stepping back, and taking stock of the complete picture. The evidence in favour of endosymbiosis had always been weaker for mitochondria than for chloroplasts, even though the main biochemical activity housed in them—respiration—is every bit as fundamental as photosynthesis. The main issue was that there was no known prokaryotic lineage to which mitochondria bore resemblance in the same way that chloroplasts were so strikingly similar to cyanobacteria. It certainly wasn't for lack of study.

In 1975, Philip John and Bob Whatley from the University of Oxford noticed that the aerobic soil bacterium *Paracoccus denitrificans* contained a particular suite of features involved in cellular respiration that resembled mitochondria more closely than did any other known bacterium.[33] Based on these features, they put forth an intriguing model for how a free-living *Paracoccus*-like cell could evolve into a mitochondrion. Their focus was on possible biochemical interactions between the endosymbiont and its eukaryotic host, and what advantage the 'protomitochondrion' might have conferred upon the cell in which it resided. But what sort of cell was *that*?

Perhaps—John, Whatley, and others suggested—it was an organism like *Pelomyxa palustris*, an obscure amoeba discovered lurking in their own back yard—more specifically, at the bottom of the stagnant, oxygen-poor 'Elephant Pond' behind Oxford's Natural History Museum in 1893.*[34] *Pelomyxa* certainly fit the bill: with its prominent nucleus it was clearly a eukaryotic cell, yet it appeared to lack other features we have come to expect in a eukaryote, most notably

---

* It was a student, Lillian Gould, who found *Pelomyxa* in the 'pond', a water-filled pit into which museum curators are purported to have dumped the carcass of an Asian elephant and left it to rot. Once suitably defleshed, the skeleton was prepared for exhibition. As of this writing the elephant is still on display.

mitochondria and an endomembrane system for shuttling proteins around the cell. Intriguingly, *Pelomyxa* made its energy (ATP) in an oxygen-independent manner, and harboured bacterial endosymbionts whose own energy needs seemed tailored to take advantage of the biochemical waste products of its host.[35] Some of these intracellular bacteria even resided within membranous compartments that were tethered to the *Pelomyxa* nucleus, which ensured that the endosymbionts were faithfully passed on to daughter amoebae after cell division. Margulis considered *Pelomyxa* and its closest amoeba brethren to be 'primitive anaerobic species', which could be living descendants of the 'protoeukaryotes'.[36]

The information gleaned from *Paracoccus* underscored the value of thinking 'outside the box', of moving beyond the constraints imposed by the handful of 'model' species that most biologists paid attention to. According to John, '*Paracoccus* was a much better reference point for considering the evolution of the mitochondrion than was *E. coli*'.[37] And, as we shall learn in Chapter 6, mitochondrion-lacking eukaryotes such as *Pelomyxa* presented exciting opportunities for testing hypotheses about early events in the evolution of the eukaryotic cell. Nevertheless, as interesting as they were, John and Whatley's data supporting a *Paracoccus*-mitochondrion connection did not strongly favour the xenogenous scenario for the origin of mitochondria over autogeny. The fact remained that there were no clear predictions as to what rRNA sequence data might reveal about the origin of this quintessentially eukaryotic organelle.

Of course, none of this mattered if those data couldn't be collected. Compared to chloroplasts, mitochondria were an experimental nightmare. Their rRNAs were present at a much lower abundance in the cell, and would be that much harder to isolate and sequence. It was Gray's choice of experimental system that made it possible.

The journey began with wheat germ—the embryos of wheat seed that, under the right conditions, germinate to produce new plants. Using commercially available material Gray's team were successful in figuring out how to purify mitochondria, a process that involved breaking the cells open, spinning the resulting cell slurry in a centrifuge, and isolating the specific sub-cellular fraction that contained the organelle and the nucleic acids within. But there was a problem. The commercially produced embryos were damaged goods; they were non-viable. They would not germinate and thus could not be used as the starting point for producing radioactively labelled RNA.

They would need to make their own wheat germ. Gray's student, Scott Cunningham, was enthusiastic about the prospects from the beginning, Gray himself considerably less so: 'Our calculations showed that we would have to isolate gram quantities of viable embryos (by grinding up wheat seeds and extracting the embryos, a laborious procedure) and incubate them with huge quantities of radioactivity'.[38] It certainly wasn't very safe, was it even possible? In the end Gray's curiosity got the better of him; he gave Cunningham permission to try and find out.

A wide range of 'scientific' equipment was assembled for the task. In addition to the standard lab fare—centrifuges, beakers, and so on—this included a jumbo kitchen blender, sieves, and a hair dryer. A whopping 16 grams of wheat embryos were isolated and germinated in the presence of radioactive phosphate; as they prepared to spring into life, the embryos began synthesizing copious amounts of rRNA, which became heavily labelled with radioactivity. Mitochondria were then purified and the radioactive ('hot') rRNA was extracted and digested. As one might expect, the labware became hopelessly contaminated: 'We worked behind Plexiglass shields and monitored everything with a Geiger counter—ridiculous, really, because everything was so hot that the counter was always off-scale'. The offending flasks and tubes were dumped in an empty room across the hall, where they would sit for a year or more until the radioactivity had decayed to a safe level.[39] The flammable organic solvents and high-voltage equipment used to tease the labelled RNA fragments apart from one another were even more dangerous. Researchers carried out their experiments in a dedicated 'electrophoresis room' equipped with a CO_2 emergency system in case of fire.[40]

Safety concerns aside, what mattered was the end product: suffi-cient rRNA was obtained, and both the mitochondrial and cytosolic rRNAs were catalogued. It was Bonen who led the final assault, translating the X-ray spots into rRNA sequences. Bonen, Cunning-ham, Gray, and Doolittle published their results together, in 1977, concluding that wheat mitochondrial rRNA is demonstrably prokary-otic in nature.[41] As was the case for chloroplasts, the similarities between the mitochondrial and cytosolic rRNAs were extremely low. While the data were consistent with an endosymbiotic origin for mitochondria, it wasn't possible to point to a specific bacterial lineage as the probable progenitor of the mitochondrion—that would come later. But for the moment it didn't matter. Molecular phyloge-netics seemed to be living up to its promise as the ultimate tool with which to infer the history of life.

Phylogenetics meets philosophy

Evolutionary speculation constitutes a kind of metascience, which has the same intellectual fascination for some biologists that metaphysical speculation possessed for medieval scholastics. It can be considered a relatively harmless habit, like eating peanuts, unless it assumes the form of an obsession; then it becomes a vice.

Roger Stanier, 1970[42]

In 1980 a symposium entitled Origins and Evolution of Eukaryotic Intracellular Organelles was held in New York City. The attendee list reads like a who's-who of the field. Many of the researchers we have discussed in this chapter were there, presenting their newest results and debating the issues. Tough questions were asked; thoughtful answers were given. There was shouting involved. Collectively they were trying to wrap their brains around the morphological, biochem-ical, and molecular sequence information that now lay before them. At the end of the meeting and out of simple curiosity, Doolittle suggested that they put it to a vote—who in attendance thought that mitochondria and chloroplasts were of endosymbiotic origin? It is unclear whether a vote actually took place, but Bonen distinctly remembers the strenuous objections of Lynn Margulis: 'That's not how science works!'[43]

Judging from the tenor of his opening remarks, the symposium organizer Jerome Fredrick (1926–1995) appears to have been anticipating heated arguments. Fredrick, a biochemist at New York's Dodge Chemical Company, spoke of Stanier and his musings about peanuts and evolution. He gently reminded the delegates of what he saw as the fine line between 'speculation' and 'dogma':

> Whether we admit it or not, we have two rival schools of biology based upon evolutionary speculations...We wish to start a dialogue, and a meaningful one, between the two...However, we can only accomplish this if we admit to ourselves that neither of us is 'in possession of the absolute truth'.[44]

But what *was* the truth? Based on the collection of articles that stemmed from the symposium, it's clear that the endosymbiont hypothesis was still just that, a hypothesis that was by no means a 'done deal' in the minds of everyone. The title of the paper presented by Thomas Uzzell and Christina Spolsky spoke volumes: 'Two Data Sets: Alternative Explanations and Interpretations'.[45] The opening sentence of Cronquist's contribution made his own views perfectly clear:

> I am in basic agreement with the position of Drs. Uzzell and Spolsky that the present available data do not require acceptance of the endosymbiotic hypothesis as a general explanation for the origin of the eukaryotic condition, and that said hypothesis has some serious unresolved problems.[46]

In light of the results of rRNA phylogenies, even the most ardent sceptics could no longer deny that mitochondria and chloroplasts were of prokaryotic ancestry. But Cronquist and his fellow autogenists, most of whom were botanists, tended to see the endosymbiont hypothesis as 'all-or-nothing'—either endosymbiosis had played a role in the evolution of the eukaryotic cell or it had not. This was significant given that, as we have learned, Margulis tended to see evidence for endosymbiosis wherever she looked. Taylor had articulated both 'mild' and 'extreme' versions of SET,[47] the latter of which accommodated Margulis's hypothesis that the eukaryotic flagellum—the appendage that cells use to move around—was derived from a symbiotic spirochaete bacterium (see Chapter 3). She went as far as to refer to plant cells as being 'quadrigenomic' in nature.[48] It was a spectacularly bold statement given that at the time there was no evidence for a fourth genome! Nor has there ever been. Despite decades of searching and a few unfortunate false leads along the way, DNA has never been found associated with the flagellar apparatus of a eukaryotic organism.[49] It proved to be the genome that never was.

The case for a spirochaete-flagellum symbiotic connection was extremely weak, and in the eyes of some served to undermine the validity of the data for mitochondria and chloroplasts. Yet other researchers saw no reason to throw the baby out with the bathwater. What was undeniable, and what Klein, Cronquist, and their supporters tended to ignore, was the power of the chloroplast and mitochondrial rRNA data when considered together. In isolation, each of the datasets could be interpreted within the framework of autogeny. Perhaps, it was argued, these organelles were sub-cellular compartments that had somehow managed to hive off a bit of DNA from the nucleus, whose own ancestry could be traced directly back to a prokaryotic cell. Endosymbiotic mergers need not be invoked.

Doolittle, Gray, and the other molecular biologists acknowledged that this scenario could explain the evolutionary origin of chloroplasts *or* mitochondria. But because of the distinctive nature of their respective rRNA sequences, it was impossible for *both* organelles to be autogenously derived from within the same cell. Their rRNAs were both prokaryotic, but they were not specifically related to one another. Simply put, mitochondria and chloroplasts could not have both descended from the same prokaryotic stock that ultimately gave rise to the eukaryotic nucleus. So if mitochondria *or* chloroplasts had evolved by endosymbiosis, then why not both? As more and more molecular data accumulated, it became clear that no other explanation would suffice.

Has the endosymbiont hypothesis been proven?

Not long after the dust had settled from the initial flurry of molecular tests of the origins of mitochondria and chloroplasts, Gray and Doolittle published a review article whose title posed this very question. It was an exhaustive (and exhausting) document—42 pages in length and with 488 references to scientific studies, it summarized a mountain of data.[50] It was 1982 and the short answer was 'yes' for chloroplasts and 'no' for mitochondria. The complete sequence of the maize chloroplast 16S rRNA gene had been determined by Zsuzsanna Schwarz-Sommer (1946–2009) and Hans Kössel (1934–1995) of the University of Freiburg, Germany,[51] analysis of which further bolstered the case for its prokaryotic origin; knowledge of the similarities between the molecular biology of chloroplasts and cyanobacteria was growing by leaps and bounds, thanks to researchers such as Harvard University's Lawrence Bogorad (1921–2003), Düsseldorf's Reinhold Herrmann, and John Gray at Cambridge.[52] Of the doubts that remained, most were related to the fact that mitochondria had yet to find a clear home on the rRNA-based phylogenetic tree of prokaryotes. Until that happened, and until the evolutionary origin of the eukaryotic nucleus was better understood, Gray and Doolittle were unwilling to consider the case closed, at least in print. They didn't have to wait for long.

In 1985 the Woese lab used their latest haul of rRNA sequences to great effect, showing that mitochondrial rRNAs from diverse eukaryotes, including wheat, fungi, and mice, were most similar to a newly obtained sequence from a prokaryotic plant pathogen by the name of *Agrobacterium tumefaciens*. The significance of this evolutionary affiliation was not lost on the authors:

The endosymbiont that gave rise to the mitochondrion belonged to the α subdivision [of bacteria], a group that also contains the rhizobacteria, the agrobacteria, and the rickettsias—all prokaryotes that have developed intracellular or other close relationships with eukaryotic cells.[53]

If the 'rickettsias' sound familiar, it's probably because they include organisms like *Rickettsia prowazekii*, an intracellular parasite that has caused more than its fair share of misery for *Homo sapiens*—it gives rise to epidemic typhus. According to Gray, *Rickettsia* and its relatives 'represent one of biology's great ironies',[54] on the one hand responsible for the death of countless millions over more than 500 years of human history, on the other hand providing key insight into the evolution of the life-giving mitochondria at the energetic heart of our cells.

RNA cataloguing had by this time gone extinct; the faster and simpler DNA-based sequencing techniques of Fred Sanger were taking hold; molecular datasets were growing exponentially. And the same wave of sequence information that had allowed the origin of mitochondria to be pinpointed signalled the end of the line for autogenous models of mitochondrial and chloroplast origins. It was now clear that the eukaryotic cell was a genetic and biochemical mosaic, a hybrid cell in which basic sub-cellular processes took place in physically separated compartments, each with a different evolutionary history, a history written in their genes. It was the start of a new era of molecular cell biological research, one freed from the constraints of 'traditional' evolutionary thinking.

I have painted a portrait of rRNA as the 'master' phylogenetic molecule, and for good reason. In 1999 Taylor considered the rRNA data in support of endosymbiosis

to have been the greatest lasting impact of molecular phylogenetics. Despite some possible changes of interpretation, principally to do with the ancestry of the eukaryotic 'host,' the acceptance of the symbiotic origin of these organelles [mitochondria and chloroplasts] constitutes a Kuhnian paradigm shift that took roughly 20 yr to complete.[55]

But we would be remiss not to consider the significance of the early protein-based studies of cellular evolution. For example, analysis of cytochrome *c* proteins by Margaret Dayhoff (the pioneering bioinformatician whom we met in Chapter 2) and her colleague Robert Schwartz were among the first to suggest that mitochondria and chloroplasts were 'bacterial'.[56] Cytochrome *c*, a protein that plays a critical role in the flow of energy through both organelles, showed much promise as a tracker of evolutionary history, but it wasn't perfect. Unlike rRNA, cytochrome *c* is not universally distributed—it does not exist in organisms that live in the absence of oxygen—and thus cannot be used to infer the evolution of all life forms.

And with respect to endosymbiosis and the origin of the eukaryotic cell, cytochrome *c* was an enigma: bacterial though it may be, the gene for mitochondrial cytochrome *c* was found to reside *in the nucleus*, not in the mitochondrion. In fact, as researchers began to study organellar genomes in greater detail, it quickly became clear that the *vast majority* of the proteins in chloroplasts and mitochondria are encoded in the nuclear genome. Your mitochondria require more than a thousand different proteins in order to keep you alive—a mere 13 of these are encoded in your mitochondrial genome. Why? The answer to this evolutionary mystery is the key to understanding how endosymbionts become organelles. It also provides interesting clues as to how prokaryotes might have become eukaryotes—how complex life evolved.

6

THE COMPLEX CELL

When, Who, Where, and How?

[A]ny living cell carries with it the experiences of a billion years of experimentation by its ancestors. You cannot expect to explain so wise an old bird in a few simple words.

Max Delbrück, 1949.[1]

The lowly amoeba commands little in the way of respect. Easy to draw and even easier to poke fun at, it seems to get as much attention from cartoonists as it does from scientists.* Which is a shame, because these 'simple', 'shapeless', single-celled eukaryotes are far from the boring sacks of jelly they are usually made out to be. Just ask Kwang Jeon, Emeritus Professor of Biochemistry at the University of Tennessee at Knoxville. Jeon has spent his entire career studying the secret lives of amoebae. His personal favourite is a giant amoeba that goes by the name of *Amoeba proteus*, 'giant' because at up to 0.5 millimetres in length, it can be seen with the naked eye. Jeon has studied where *Amoeba proteus* lives, how it moves, how it eats, and what it eats. He knows it inside and out—literally.

In 1966, early in his career, one of Jeon's precious amoeba strains came down with a bacterial infection. Amoebae generally don't have a problem with bacteria—they are quite happy to eat them for breakfast. But this particular situation seemed dire, for the amoebae at least;

---

* A cartoon in *The Far Side* series by the American Gary Larson depicts what appears to be a couple of amoebae wearing glasses, lounging in chairs, and having an argument. The beer-drinking one says: 'Hey! I got news for you, sweetheart!...I am the lowest form of life on earth!'

tens of thousands of rod-shaped bacteria were having a field day inside each and every amoeba cell. Jeon could have simply discarded the culture and focused his attention elsewhere. But he didn't. He was intrigued: where had these bacteria come from and what were they doing in there? He decided to watch what happened—he watched *for a very long time.*

Most of Jeon's infected amoebae died. Those that didn't grew extremely slowly. But as the months and then years went by a remarkable thing happened. Not only did the amoebae recover, they became completely dependent upon their bacterial 'parasites' for life.[2] And if 'set free' from the amoeba cytoplasm the bacteria themselves would not grow. Equipped with what can only be described as an endless supply of patience, Jeon performed microsurgery on the amoebae, swapping the nuclei of infected and uninfected cells. Nuclei taken from the cytoplasm of infected amoebae were found to be incapable of supporting the growth of cells into which they were injected unless the bacteria were also present.

The precise nature of the addiction of *Amoeba proteus* to the so-called X-bacteria was never determined. It was an exceedingly complex system to work with: *Amoeba proteus* cells are in fact host to a variety of different microbes, both transient and permanent, and it was difficult to figure out who was influencing whom. Nevertheless, Jeon and colleagues did manage to gain insight into the physiological interactions between the amoeba and its newest tenant. The X-bacteria had begun to manipulate the functioning of certain genes in the amoeba nucleus; the amoeba had done the same in reverse; and the X-bacteria that survived avoided being chewed up and eaten by sheltering themselves in a digestion-resistant, membrane-bound compartment. They showed that it could take as little as 18 months—approximately 200 cell generations in amoeba-time—for the inter-dependence between the amoebae and the X-bacteria to become irreversible. It was far from obvious who was leading, but a complex genetic, biochemical, and cell biological dance was clearly taking place between the two partners.[3]

The early supporters of the endosymbiont hypothesis were keen on Jeon's research, and rightly so. It was impressive stuff and the temptation to extrapolate from it backwards in time was irresistible. If obligate, undoable relationships between eukaryotes and prokaryotes could form over human timescales in the lab, then surely a bacterium could evolve into a mitochondrion over many millions of years. This sounded reasonable enough in the 1970s and 1980s, and it still does today—but only if one doesn't think too hard about it. It's a classic problem in evolutionary biology, one that we have touched upon already: with any given system of study, how confident can we be that the processes we are observing actually resemble those that happened in the ancient past?

We now know a great deal about the molecular sequence of events that act to cement the relationship between endosymbiotic partners in nature. One key element is the movement of genes from the genome of the endosymbiont to that of the host cell. As we shall learn, not only is 'endosymbiotic gene transfer' something we can observe in 'real time', genomic investigations of mitochondrial, chloroplast, and nuclear genomes of diverse eukaryotic organisms allow us to be very confident that it is something that has happened a lot throughout the history of life. And yet when it comes to the 'when, who, where, and how' of eukaryogenesis there are still a great many uncertainties. The more we learn the more we still need to figure out.

In this chapter we will investigate the role of endosymbiosis in the origin of the eukaryotic cell. In doing so, we can safely set aside the question of chloroplast evolution, for the following reason: it is very clear that the organism that played host to the cyanobacterium that became the chloroplast was a eukaryote. It was a complex cell: it had all the cranks, levers, and buttons we've talked about, a nucleus, cytoskeleton, intracellular protein transport system, and so on. The cyanobacterium was taken up by this eukaryote, probably in the same way (and using the same basic cellular machinery) that an amoeba gobbles up bacteria from its surroundings. Because of the free energy (in the form of sugars derived from photosynthesis) that it so readily produced, the cyanobacterium must have been of significant benefit to its host, and over time it became a permanent sub-cellular fixture. From that original evolutionary dance stemmed the very first photo-synthetic eukaryotes, the first algae that developed and diversified into the myriad of unicellular and multicellular forms we see around us. We will hear all about it in Chapter 7, including a few unexpected twists and turns that happened along the way.

This general scenario—eukaryote ingests prokaryote, prokaryote gets comfortable, prokaryote becomes organelle—could well have played out and given rise to the mitochondrion, and for a time it seemed likely that it had. But exciting new possibilities have presented themselves. Researchers are still sorting out what the primordial cell that played host to the bacterial (more specifically, α-proteobacterial) progenitor of the mitochondrion was like. Was it a eukaryote or something very like it? Or was it nothing of the sort? Let's explore how the complex cell evolved, how our ancestors evolved. It's helpful to start from the beginning.

Oxygen—lifeblood or toxic waste?

The universe is approximately 14 billion years old. About 4.5 billion years ago, hot on the heels of the massive supernova explosion that gave rise to our solar system, planet Earth was formed from the agglutination of stardust: a massive sea of dust that was swirling around our newly minted sun, dust that contained the raw ingredients for life. Less than a billion years after that—after the Earth's surface had cooled and the oceans had formed—the first self-replicating 'life forms' came into existence. At first these were probably just molecules or small collections thereof, perhaps short RNA molecules that could make copies of themselves from nucleic acid building blocks, as certain types of RNA (but not DNA) are capable of doing today.[4]

As the very first membrane-bound cells evolved, a 'genetic code' was established, *the* genetic code, the one that today links DNA to protein, genotype to phenotype. As discussed in Chapter 2, the universality of the genetic code provides strong evidence in support of the idea that all known life forms stem from a common ancestor. What exactly that common ancestor was like, and how 'cellular' it was, is not clear. But it was certainly a 'simple' organism by any reasonable approximation.

Various lines of geological and chemical evidence point to a distinct, biologically induced rise in the level of oxygen on Earth beginning around 2.4 billion years ago.[5] What this tells us is that oxygenic photosynthesis, as carried out by the ancestors of modern-day cyanobacteria (and their chloroplast derivatives) had evolved by that point. Prior to this so-called 'great oxidation event', life forms were primarily anaerobic; they eked out a living using biochemical pathways that did not involve molecular oxygen. In fact, many of these early prokaryotes would have had a serious problem with oxygen had there been much of it around in gaseous form—as molecular oxygen or O_2. Oxygen is highly reactive when not bound to other elements, and potentially catastrophic for cells that do not possess the biochemical means to deal with it.*

Opinions differ on the extent to which the great oxidation event impacted the course of evolution of life on our planet. Margulis and Sagan considered it nothing short of an 'oxygen holocaust', 'by far the greatest pollution crisis the earth has ever endured'. And it was all caused by cyanobacteria doing what they do best: absorbing photons, splitting water, making sugar, and emitting oxygen gas as a by-product. As a direct consequence of this 'sudden' rise in oxygen, Margulis and Sagan argued, mass extinctions occurred. The result was a major overhaul of the microbial biosphere.[6]

Maybe this happened and maybe it didn't. Researchers such as Nick Lane, a biochemist at University College London, take a decidedly

---

* When metabolized by living organisms, oxygen can give rise to so-called 'reactive oxygen species' which hurtle around the cell at tremendous speed, damaging DNA and anything else they collide with. Organisms that require oxygen, or that can at least tolerate it, possess proteins that act to 'neutralize' such species. Small molecule antioxidants such as vitamin C act similarly.

more neutral position on the role of oxygen in the evolution of life—toxic, yes, but lifeblood too.[7] And just because there were cyanobacteria around belching out oxygen doesn't necessarily mean that it was accumulating in the atmosphere to any significant degree. Most freshly produced oxygen would initially have been chemically 'captured' by dissolved iron and organic particles in the oceans, a process referred to as oxidation (this is what causes cars to rust in the presence of moisture). It was only after the various 'oxygen sinks' were full that levels of free oxygen gas would have actually spiked. Precisely when this occurred is unclear.

Regardless, it seems reasonable to conclude that an increase in molecular oxygen levels, in the atmosphere and in the oceans, would have had a significant impact on where and how cells could grow. Organisms that (for whatever reason) did not evolve ways of dealing with its damaging effects would have found themselves increasingly isolated, restricted to anoxic or low-oxygen environments in which they would not be poisoned. (Note that there are many such places on today's Earth: marine sediments, in the guts of animals, and so on.) And many of the organisms that *did* learn to live with oxygen would have taken advantage of it, spreading into new, increasingly oxygen-rich niches as they became available.

Mitochondrial metabolism: what's oxygen got to do with it?

Before proceeding further on our journey towards the complex cell, let's pause for a moment and consider why oxygen matters to us. Human beings depend on oxygen for life because our mitochondria cannot make ATP without it. Which, of course, explains our long-standing obsession with all things mitochondrial.

The biochemical connections between the food we ingest, the air we breathe, and ATP production are complex; it took researchers a long time to get it right. It was like a maze in which the entrance and exits were obvious—start with glucose, end up with ATP—but the precise path between them was unclear. There were rules to follow, educated guesses as to which chemical intermediates the path of energy ought to flow through, based on decades of experimental data. But by the mid-1900s there were still gaps.

In 1961, an eccentric British biochemist named Peter Mitchell (1920–1992) put forth a radical hypothesis that, in principle, completed the maze. Mitchell's hypothesis provided a complete account of the 'checks and balances' of ATP synthesis. But it was, as we might say today, outside the box; it was far too radical for most biochemists to stomach. Consequently, Mitchell was scorned for many years. But he was (in most respects) right. He received a Nobel Prize in 1978 for elucidating the role of the 'bioenergetic membrane' in cellular metabolism.[8] The implications of Mitchell's discovery extend far beyond our own human-centric view of mitochondrial ATP synthesis.

What Mitchell hypothesized, and what was ultimately proven experimentally, was that it is the double-membrane envelope that surrounds the mitochondrion, and the sea of proteins embedded within it, that is the key to high-efficiency ATP synthesis.[9] As the appropriate biochemical precursors (derived from the breakdown of energy-rich glucose in the cytoplasm) are fed into the mitochondrion, they trigger a cycle of chemical reactions, the end result of which is that hydrogen atoms—positively charged 'protons' (H^+)—are 'pushed' through a series of one-way channels into the tiny space between the two membranes. This unidirectional flow results in a higher concentration of protons on one side of the membrane than on the other. It is called an electrochemical gradient. The protons then tumble back through another one-way channel by simple osmosis—as sure as gravity drives sand grains through an hourglass.

Wait, why bother to push protons across a membrane only to have them come straight back again? And what does oxygen have to do with it? The one-way channel through which protons move from an area of high concentration to one of low concentration is actually a highly sophisticated ATP synthesis 'factory'. As long as protons continue to be pumped into the intermembrane space, a proton gradient is maintained and protons are 'compelled' to move back to the side where the concentration of protons is lower—and the so-called 'ATP synthase' keeps making ATP (Fig. 4). In humans, for every three protons that pass through the synthase channel, one molecule of ATP is synthesized. And it is oxygen that plays a critical role in the chemical reactions that drive the pumping of protons 'up-hill' so that they can come back down again. Without molecular oxygen, the proton pumping chain grinds to a halt and so too does ATP production.*

The biological significance of membranes cannot be overstated. Although the details vary from lineage to lineage, Mitchell's general principle of 'chemiosmosis'—the harnessing of an electrochemical concentration gradient across a membrane—underlies the process of

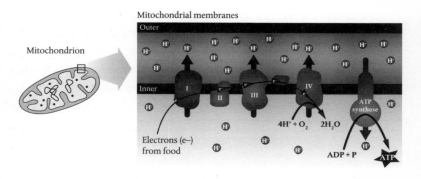

Fig. 4. Simplified schematic diagram showing the process of ATP synthesis in oxygen-utilizing mitochondria. Electrons (e−) derived from the chemical breakdown of food (e.g. glucose) are passed down the so-called 'electron transport chain', of which there are four complexes (I–IV). Energy released during the flow of electrons down the chain drives the pumping of protons (H^+) across the inner mitochondrial membrane by complexes I, III, and IV. The unidirectional movement of protons generates an electrochemical gradient—a higher concentration of protons in the space between the inner and outer mitochondrial membranes than in the main mitochondrial compartment. Protons then move back through the membrane from an area of high to low concentration via the ATP synthase channel, a 'molecular motor' that produces ATP from the addition of a phosphate (P) to ADP. Assuming a continuous influx of electrons and oxygen, ATP will continue to be synthesized. In human beings, one molecule of ATP is synthesized for every three protons that pass through the channel.

ATP synthesis across vast swaths of life. It is emphatically *not* a mitochondrion-specific phenomenon; it occurs in many prokaryotes as well (in this case protons are pumped into the narrow space between the pair of membranes that surround the cell). And chemiosmosis occurs in chloroplasts, where, as in cyanobacteria, it is involved in the first step of photosynthesis where proton pumping is triggered by the absorption of photons by membrane-associated pigments and proteins. Part physics, part chemistry, the evolution of the bioenergetic membrane appears to have gone hand in hand with the evolution of life and, in particular, complex life.[10] We shall return to this critical point at the end of the chapter.

Given the essential role for oxygen in the metabolism of our mitochondria, it is perhaps not surprising that oxygen tends to feature prominently in models of eukaryotic evolution. Such models often revolve around the question of when oxygen-utilizing (not oxygen-producing) prokaryotes evolved and, in particular, how—and why—this dependence on oxygen made its way into the eukaryotic domain. With this in mind, let's take a deep breath and rejoin the path to eukaryotes.

The complex cell: when?

The use of fossils to track the evolution of microbial life is a tricky business. As discussed in Chapter 4, it is one of the main reasons why molecular sequence-based approaches are so appealing. Nevertheless, there are several landmarks that have been established from the study of 'microfossils' that bear on the question of early cell evolution, and

---

* The proton pumping chain is more accurately referred to as an 'electron transport chain'. This is because it is the flow of electrons from one protein-containing channel to the next that drives the movement of protons from one side of the membrane to the other (Fig. 4). The electrons are derived from food such as glucose. Note also that this membrane-based ATP synthesis process isn't the only way ATP is made, but it is far and away the most efficient; in aerobic organisms such as ourselves, it accounts for the vast majority of ATP underlying cellular metabolism.

eukaryotic evolution in particular. These fossil-based waypoints serve not as hard dates but as soft bounds on the problem, to be considered with reference to molecular data and biogeochemical information. For example, if the eukaryotic cell was aerobic from the very beginning, when do eukaryotes first appear in the fossil record and how does this appearance square with the timing of the great oxidation event?

The earliest fossils generally considered by scientists to be derived from living cells are astonishingly old: *3.5 billion years old*—about three-quarters the age of the Earth. These fossils are believed to represent 'cyanobacteria', the organisms that catalysed the great oxidation event more than two billion years ago. I say 'cyanobacteria' here because in the case of such ancient fossils, it is generally not the cells themselves that fossilize but the structures they give rise to. A trademark of the growth of cyanobacteria in shallow water is the formation of *stromatolites*: stratified, columnar structures that are comprised of layer upon layer of microbial mats mixed with sedimentary particles (the word stromatolite roughly means 'layered rock'). The cyanobacteria are at the top—to avoid being covered over by sediment, they grow steadily upward in their never-ending quest for light, adding new layers to the inorganic (and fossilizable) components of the stromatolite underneath.[11]

Modern-day stromatolites aren't very common, but they do exist—this is how we know what they are comprised of in the first place. Shark Bay, Australia, for example, harbours massive stromatolite beds in which the individual columns can be up to a metre high. They grow upwards at a rate of about 5 centimetres every 100 years. Stromatolites appear to have been abundant along the seashores of the ancient Earth. It is unclear how 'cyanobacterial' the microbes that produced the oldest known stromatolites were. Were they the true ancestors of modern-day cyanobacteria, or were they something completely different, another type of bacterium with a similar growth habit? And, strictly speaking, we cannot be certain that 3.5 billion-year-old stromatolites were the product of growing microorganisms at all. Some researchers, such as Oxford palaeontologist Martin Brasier, have argued that some stromatolites are derived from geological processes, not biological ones. Nevertheless, there is robust evidence for stromatolite-associated fossilized prokaryotes stretching back an impressive 2.7 billion years into the past.[12] It seems very likely that these organisms were carrying out oxygenic photosynthesis.

Within the eukaryotic domain, microfossils of various shapes and sizes are reasonably common in rock samples taken from around the globe dating back half a billion or so years. Among the most impressive are a series of vase-shaped fossils discovered in rock from Arizona's Grand Canyon, which look remarkably like what today are called testate amoebae—amoeba cells surrounded by distinctive shells, or 'tests' (it is the tests that mineralize and can be fossilized). These fossils are ~0.75 billion years old.[13]

More impressive still is a diverse group of putatively eukaryotic fossils that some palaeontologists call 'acritarchs'—impressively ambiguous that is. The term 'acritarch' means 'of undecided origin'. Acritarchs range in size from ~5 to 200 micrometres in diameter—vastly larger than the majority of known prokaryotic cells—and some even approach 1 millimetre in length. They come in a plethora of different shapes: many are oval with radially protruding stubby flagellum-like appendages; others are vaguely reminiscent of modern-day worms. Based on their size and apparent morphological complexity, it's tempting to conclude that acritarchs are eukaryotes, but the fact remains that they could be just about anything. Some researchers believe that acritarchs represent early eukaryotic lineages that went extinct. Perhaps they correspond to 'transitional forms', somewhere between prokaryotic and eukaryotic cells. Whatever they are, they are truly ancient, some perhaps more than three billion years old.[14]

At this point you might be wondering about the potential use of molecular data in solving the conundrum of when the earliest eukaryotic cells arose. Perhaps a 'molecular clock' approach might work? If, for instance, one were to compare the amino acid sequences of proteins encoded by the genomes of both prokaryotic and eukaryotic cells, and one knew how fast the molecular clock 'ticked', one could tally up the observed protein sequence differences between them and compute a divergence time in 'X' billion years. Sounds easy, right? Researchers have tackled the problem using precisely this approach, most notably the American biochemist Russell Doolittle of the University of California, San Diego.[15] The results were problematic to say the least.

In the mid-1990s, Doolittle and colleagues assembled a large dataset of protein sequences from organisms spanning the full breadth of life's diversity—bacteria, archaea, and multicellular and unicellular eukaryotes. Having calculated the amino acid sequence differences within and between the major lineages represented, they used a molecular clock calibrated from a set of well-studied animal fossils to extrapolate backwards in time. They estimated that eukaryotes and prokaryotes shared a common ancestor approximately two billion years ago.[16] Really? *Two* billion years ago? This wasn't nearly long enough given that cyanobacteria—a distinct lineage within the bacteria—were supposedly already on the Earth 3.5 billion years in the past. Other researchers, analysing the same data as Doolittle's team but with different technical procedures and assumptions, came up with much older divergence time estimates, anywhere between 3.5 billion and six billion years ago![17] Such wild (and in some cases literally impossible) estimates prompted the evolutionary biochemist William Martin to ponder: 'Is something wrong with the tree of life?'[18]

The answer to Martin's rhetorical question, as you will have guessed, was 'yes', and we are still trying to figure out precisely what that 'something' is. But we have learned a few important lessons along the way. Molecules do indeed exhibit clock-like behaviour over 'short' evolutionary timescales, and molecular clock analyses have proven extremely valuable in diverse areas of biology. For example, they have been used to track viral epidemics and to time the diversification of certain groups of animals, to name but two successful applications. However, molecular clocks become terribly unreliable when pushed too far. And knowing how far is *too* far is not something that is written in stone. What we can say with certainty from years of painstaking investigation is that the rate at which DNA and protein sequences change over time can speed up and slow down in different groups of organisms; it is extremely difficult to determine whose molecular clock has decided to tap out a different rhythm.[19] So what good is a clock that doesn't keep accurate time? It's better than no clock at all. A watch that loses a minute on one day and gains two minutes the next will still get you to work more or less on time, but it would be unwise to rely upon it without having it calibrated on a regular basis. And so it is with molecular evolution.

You may have thought of another factor that has the potential to confound molecular clock-based inferences of ancient cellular divergences: endosymbiosis. Endosymbiosis serves to bring evolutionarily distinct organisms—and their genes—together into the same organism. Consequently, even with a perfectly reliable molecular clock, divergence time estimates will vary significantly depending on which genes are selected for analysis, because all of the genes present in the organism do not trace back to the same cellular ancestor. Indeed, as we have learned more and more about the evolutionary history of the genetic material housed in the eukaryotic nucleus, scientists have concluded that this is precisely the issue. The nuclear genome is a mosaic of DNA from two evolutionarily distinct sources: bacteria and archaea. How this came to be is what evolutionary biologists wish to understand.

Let's pause once more and take stock of what we've learned about when the eukaryotic cell evolved. The fossil evidence for the earliest eukaryotic cells is inconclusive, but it is safe to say that bona fide eukaryotes go a long way back, at least three-quarters of a billion years. Molecular clock estimates for the age of the prokaryote–eukaryote divergence are controversial and should be taken with a grain of salt. We now shift our focus away from the vexing question of timing and towards the more fundamental issue of mechanism. Yes, we would like to know when the eukaryotic cell evolved. But having an answer to that question would be immensely unsatisfying without knowing *how* it happened. And for that we return to where this chapter began, to the realm of cell biology.

Eukaryotes without mitochondria (?)

Tom Cavalier-Smith is a singularity in the field of cell evolution. 'Have you seen my latest paper on phagocytosis?' he asks. 'Er, which one?' I stammer. I could be forgiven for being uncertain; his reputation for churning out manuscripts the size of bibles is legendary, and he does so like clockwork. Agreeing to review a Cavalier-Smith paper usually comes with a sinking feeling of overcommitment; do I *really* have time for this? And will I even be able to understand it? It is a rare biologist who makes it their business to consider information gleaned from so many disciplines—from genetics to geology, biophysics to bioinformatics. Margulis was one. Cavalier-Smith is another.

Professor Emeritus of Evolutionary Biology at the University of Oxford, Cavalier-Smith received classical training in zoology at Cambridge University in the 1960s. After a stint as Reader in Biophysics at King's College, London, and as Professor of Botany at the University of British Columbia, he returned to the UK in 1999. He has been a member of Oxford's Department of Zoology ever since. Cavalier-Smith is a proponent of what he calls '3D cell biology': he seeks to understand the evolution of cells by studying the relationship between cellular structure and function. He speaks of cellular 'body plans', which is to say that he thinks about cell morphology in the same way a palaeontologist thinks about the skeletal system of a dinosaur. He studies cells in a holistic fashion, as a physiologist considers how tissues and organs work together to sustain the growth and reproduction of an animal. And he thinks about membranes—a lot: how they act to partition cells into discrete, functionally distinct compartments, and how membrane-associated proteins carry out their jobs in diverse life forms.[20]

Although not content to focus on particular groups of organisms, Cavalier-Smith does have his favourites. In the 2000s, for example, he was particularly fascinated with the hidden diversity of microbes inhabiting soil. And as we will learn in Chapter 7, he has a long-standing interest in photosynthetic lineages. But he knows as much about birds and butterflies as he does about bacteria and beech trees. Over the past 40 years his mission has been to assemble a hierarchical classification scheme that accounts for absolutely every living thing. In this sense he is something of a modern-day Carl Linnaeus (1707–1778), the Swedish botanist considered to be the father of modern taxonomy. Cavalier-Smith is a hypothesis-generating machine and he is not afraid to change his mind. It drives many biologists crazy.

Cavalier-Smith's bread and butter is 'protistology'—the study of everything with a nucleus that isn't an animal, fungus, or plant. If that strikes you as an odd definition and an ambitious task, you are not alone. The 'protists' (or protozoans) are a loosely defined grab bag of eukaryotes, exceptionally diverse in every sense of the word. The vast majority of protists are single-celled, and it is for this reason that they occupy such a special place in the hearts and minds of researchers who study the evolution of eukaryotes. It is from within the protists that the macroscopic life forms that surround us today sprang. And the protists have been an invaluable source of clues as to how the very first eukaryotic cell, and its mitochondrion, must have evolved.

In the 1980s Cavalier-Smith articulated a provocative hypothesis about early eukaryotic evolution that fuelled research in the field for more than a decade. It was grounded on his belief that the advent of phagocytosis—'cell eating'—was a critical step in the development of the complex cell, one that researchers tended to ignore. Lynn Margulis, for example. In her original formulation of eukaryotic cell evolution, Margulis presented both the host cell and the protomitochondrial endosymbiont as being prokaryotic in nature,[21] but gave relatively little thought to the problem of *how* one cell had come to reside within the other. Other early proponents of endosymbiosis, however, such as Roger Stanier (1916–1982) and the Nobelist Christian de Duve (1917–2013), felt the issue of phagocytosis was simply too important to be overlooked.

The cytoskeleton is the complex internal scaffolding that gives the eukaryotic cell its shape. It plays an essential role in the process by which all eukaryotic cells ingest foreign matter; human cells such as 'macrophages', for example, patrol our bodies, gobbling up potential pathogens and cancer cells by phagocytosis in a manner indistinguishable from that exhibited by an amoeba in search of a meal. In the 1960s and 1970s, de Duve and Stanier had argued that the ability to take up organisms (as food) from the surrounding environment would have been of considerable benefit to the pre-mitochondrial ancestors of the eukaryotic cell.[22]

Inclined to think of cell evolution in mechanistic terms, Cavalier-Smith agreed. He proposed the existence of a group of unicellular eukaryotes called *archezoa*—'ancient animals'.* The archezoa were purported to be relics of a critical phase of eukaryotic evolution: they had branched away from the main trunk of the eukaryotic tree after the cytoskeleton had evolved but *before* the endosymbiotic origin of mitochondria.[23] If the 'missing links' leading to eukaryotes could not be found in the fossil record, perhaps they were still among the living? Like all good scientific hypotheses, Cavalier-Smith's archezoa hypothesis made explicit predictions that could be tested. The results were to trigger a sea change in our understanding of the diversification of eukaryotic cells.

We have already met one such 'archezoan' in Chapter 5: *Pelomyxa palustris*, the anaerobic amoeba discovered by Lillian Gould in the Elephant Pond behind the Natural History Museum in Oxford. As eukaryotic cells go, *Pelomyxa* appeared to be 'simple'—and it seemed to lack mitochondria. In 1973, Bovee and Jahn had proposed that *Pelomyxa* might be a 'transitional eukaryote',[24] and Cavalier-Smith added a variety of other simple-looking eukaryotes to the list of potentially primitive organisms. These included mitochondrion-lacking eukaryotes that lived in the absence of oxygen (or could tolerate only small amounts of it), such as *Giardia lamblia* (the cause of 'beaver fever', an unfortunate diarrhoea contracted by hikers who unwittingly ingest its cysts from stream water); *Trichomonas vaginalis* (which causes sexually

---

* The archezoa are not be confused with archaea (or archaebacteria), the latter being the highly distinctive lineage of prokaryotes discovered by Carl Woese.

transmitted disease in humans); *Vairimorpha necatrix* (a troublesome pathogen of moths); and *Entamoeba histolytica* (causative agent of amoebic dysentery). From this brief (and somewhat depressing) list of protists you will have no doubt detected a trend: most of the organisms thought to be archezoans were parasites.

What awaited Cavalier-Smith's archezoa was nothing less than a full-on assault on virtually every aspect of their biology. Their fine-scale structural features were investigated using the electron microscope; biochemical studies were carried out to try to understand their metabolism; and, of course, their genes were sequenced. It was the 1980s and ribosomal RNA (rRNA) had proven its worth as a tool for inferring the evolutionary history of anything and everything microbial. All that remained was for it to be applied to the realm of protists.

The first such rRNA-based phylogenies were exciting to say the least. The mitochondrion-lacking archezoan lineages were found to be vastly different from all other eukaryotes for which sequences were available for study (Fig. 5).[25] The nucleotide sequence of their rRNAs, when lined up against those of plants, animals, and fungi, were like nothing scientists had ever seen. The evidence in hand was consistent with Cavalier-Smith's hypothesis: the sequences from mitochondrion-lacking lineages branched off at the very base of the tree, distinct from those derived from eukaryotes that harboured mitochondria. The excitement over what these organisms might represent—bona fide transitional eukaryotes—was palpable. It was like Woese's discovery of the archaea all over again.

Yet during the 1990s, as more and more rRNA sequences were placed on the tree, and as the computer programs used to analyse them became more sophisticated, a curious thing happened. The tree began to look less like a ladder and more like a bush. The stepwise splitting off of eukaryotic branches as one proceeded from the trunk of the tree towards the tips became rather less obvious. Gaps on the tree that had once appeared as vast evolutionary chasms were filled in, in many cases by unicellular eukaryotes that were known to possess mitochondria (Fig. 5). The archezoan taxa were highly distinct; there

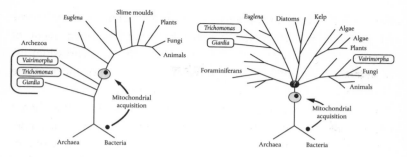

Fig. 5. Schematic evolutionary trees showing the relative positions of eukaryotic lineages with and without classical mitochondria. On the left is a stylized phylogenetic tree typical of those obtained by analysis of ribosomal RNA (rRNA) sequences in the early 1990s. Three 'archezoan' lineages—the mitochondrion-lacking eukaryotes *Giardia*, *Trichomonas*, and *Vairimorpha*—branch off the base of the eukaryotic portion of the tree. The structure of the tree was consistent with the archezoa hypothesis of Tom Cavalier-Smith, which proposed the existence of 'transitional eukaryotes', organisms that had diverged from the main line of eukaryotes prior to the endosymbiotic event that gave rise to mitochondria. As more rRNA and protein sequence data accumulated, and computational methods for building phylogenies became more accurate, the structure of the eukaryotic tree changed dramatically. The tree shown on the right illustrates the modified position of the archezoan lineages, which now branch together with mitochondrion-containing species. The '?' at the base of the eukaryotic tree illustrates present uncertainty of the deepest divisions of the tree. For a more detailed schematic of eukaryotic diversity see Fig. 10.

was no doubt about that. But on the basis of rRNA and subsequent protein phylogenetic data, the deepest branches on the eukaryotic tree all seemed to split off from one another in quick succession. By the early 2000s it had become difficult to argue, on molecular phylogenetic grounds, that the mitochondrion-lacking archezoans were any more ancient, or primitive, than were most other mitochondrion-bearing, single-celled eukaryotes.[26]

At the same time as the evolutionary history of the archezoa was being explored using phylogenetics, exciting new data began to emerge with respect to the metabolic capabilities of these anaerobic oddballs. Martin Embley, Graham Clark, Andrew Roger, Jan Tachezy, and others showed that, as remarkable as it seemed, the archezoans *did* have mitochondria—they were just unrecognizable as such. Within the eukaryotic cytoplasm, mitochondria normally stick out like a sore thumb. When viewed in cross-section they appear as sausage- or oval-shaped structures, the inner membrane of which can be seen folded back and forth on itself like pleated fabric (Figs. 1 and 4). The reason for this folding has to do maximizing the efficiency of ATP synthesis: expanding the surface area of the inner mitochondrial membrane increases the surface area across which protons can be pumped and the all-important electrochemical gradient maintained. Such inner membrane convolutions are called mitochondrial cristae; they are a hallmark of mitochondrial architecture.

In *Giardia*, *Entamoeba*, and other archezoans, no such organelle matching this description could be seen in the cytoplasm. The first evidence that mitochondria were in fact present came indirectly from the discovery of genes in their nuclear genomes that appeared to be of α-proteobacterial ancestry, the same bacterial source that gave rise to mitochondria.[27] Furthermore, these genes encoded proteins that in aerobic organisms were known to function in the mitochondrion. There was intense debate as to what these 'mitochondrial' genes were doing in mitochondrion-lacking organisms. Perhaps they had simply been acquired by phagotrophy, by the ancestors of today's archezoans ingesting bacteria and assimilating their genes.

The debate was settled when tiny membrane-bound organelles, often less than 1 micrometre in diameter, were discovered in several different archezoa and shown to be 'mitochondria'. They could not have been more different from mitochondria as described in textbooks. They had shrunken dramatically, ceased production of ATP, and lost their cristae. They had even lost the genome that 'normal' mitochondria invariably possess. Researchers refer to such organelles as 'mitochondrion-related organelles' (MROs). In the absence of a genome whose origin can be studied directly, the inference that MROs are of mitochondrial ancestry comes from evolutionary consideration of nucleus-encoded proteins that are shipped to the MRO to carry out their functions. The mechanism by which such proteins are brought inside the MRO is also distinctly 'mitochondrion-like'.

So archezoans have mitochondria, or a highly degenerate form thereof. But if they no longer make ATP for the host cell, what are these organelles doing? 'Use it or lose it' is an oft-heard phrase in evolutionary biology circles; it refers to the notion that characteristics of an organism's biology that no longer impact its ability to survive tend not to stick around for very long. In the case of archezoans, the MROs are still important; it just took a while to figure out why.

The answer was to be found among the diversity of biochemical functions that mitochondria carry out in addition to what they are best known for, oxygen-assisted ATP synthesis. This includes the synthesis of so-called iron-sulfur (Fe-S) clusters, geometrically arranged assemblages of Fe and S atoms that cells incorporate into certain types of proteins. All organisms have them; Fe-S cluster proteins play a crucial role in biochemical pathways that involve the transfer of electrons. What has emerged from consideration of diverse lines of evidence is that Fe-S cluster assembly seems to be a critical biochemical process that is carried out in all mitochondria and MROs. In the diarrhoea-causing parasite *Giardia*, making Fe-S clusters seems to be the *only* thing the MRO does. Which explains why it is all but invisible except to those who know what they are looking for. As we shall discuss in the following section, the MROs of other archezoans are somewhat less biochemically impoverished, a fact that provides important clues for interpreting the evolution of oxygen-utilizing mitochondria.

The archezoa hypothesis is dead. But the protists-formerly-known-as-archezoans have reinvented themselves like the members of a once famous but now defunct rock band. They all have stories to tell. Some, like *Giardia* and *Trichomonas*, now appear to be relatively close neighbours on the eukaryotic tree of life, branching among a slew of aerobic, mitochondrion-bearing protists. Others have gone their own way, joining up with eukaryotic lineages that are well known for other reasons. *Vairimorpha*, for example, is not the primitive eukaryote it was once thought to be—it belongs to a group of parasites called microsporidians that are degenerate fungi (Fig. 5). (You might think of the fungi themselves as 'low' forms of life, but animals share a more recent common ancestor with fungi than with plants.)

The upshot of all this evolutionary reshuffling has been a transformation in how we interpret the most obvious aspects of the biology of the supposed-archezoan protists. Their simple cellular structures are no longer considered primitive, but are best explained as a result of having adapted to specific lifestyles, in most cases a life of parasitism. As the ancestors of these unicellular creatures ventured into ecological niches containing little or no oxygen, the biochemistry of their mitochondria was radically reconfigured. In essence, they 'figured out' how to make a living by doing things differently. 'Use it or lose it' applies, but so too does 'use it, lose it, or modify it beyond recognition'. The take home message is never trust a parasite—it will deceive you every chance it can get.

The complex cell: who, where, and how?

Nucleus, endomembrane system, cytoskeleton, mitochondrion; deep in the recesses of Earth history, these sub-cellular structures came to reside within the same cell. It was a common ancestor that human beings share with all other animals—from chimps to worms, to dogs and fleas—a common ancestor we share with fungi, and with plants and algae, and with amoebae and all the other protists. It was the common ancestor of all complex life. Let's work through an evolutionary scenario for how that common ancestor might have evolved. Before beginning let me emphasize that it is but one of a variety of models for the evolution of the eukaryotic cell that have been proposed over the past 40 years, some quite similar, others radically different. We can consider it to be the 'standard' or 'traditional' model. I present it as a continuous narrative without interruption. We will discuss its merits and shortcomings relative to more modern scenarios in due course.

It all started rather modestly approximately two billion years ago within the confines of a prokaryotic cell. It wasn't a single cell, but a population of organisms that were related to modern-day archaea. These organisms made ATP in the absence of oxygen. Their DNA came to be physically separated from the rest of the cell by membranes; a nucleus was formed, probably from membranes the cell already had, perhaps from the membrane that surrounded the cell itself (Fig. 6A). The outer membrane of the nucleus was connected to, and evolved in concert with, an increasingly dynamic set of membranes capable of forming spheres—vesicles. These vesicles could fuse and split off from one another like soap bubbles; they were the seeds of the endomembrane system, what would become a highly efficient intracellular cargo transport system.

As the cells became increasingly membranous and compartmentalized, a protein-based internal scaffolding system developed. This early cytoskeleton served to connect different areas of the cell, a sea of railway tracks that was used to direct the movement of cargo in three dimensions. The cytoskeleton also allowed the cells to become larger and to change their shape in response to the environment. Cell movement was greatly enhanced by the evolution of flagella, whip-like appendages made of cytoskeletal proteins. Together with the nascent endomembrane system, the evolving cytoskeleton allowed for the development of phagocytosis, a process whereby portions of the outer cell membrane could be internalized, bringing extracellular matter inside the cell.

At some point the descendants of these anaerobic cells, which were now in possession of many of the internal features we associate with eukaryotes, found themselves in the vicinity of another group of prokaryotes: α-proteobacteria. These α-proteobacteria were able to deal with oxygen; they were producing ATP by cellular 'respiration'. By this time the 'protoeukaryotes' had been flexing their newly trained phagocytotic muscles—they were no longer simply passively absorbing the organic material needed to fuel their metabolism, they were gulping it down. The α-proteobacterial cells were ingested and

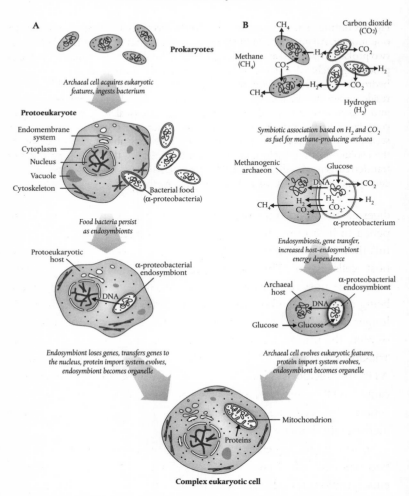

Fig. 6. Two evolutionary scenarios for the origin of the mitochondrion and the eukaryotic cell. The traditional view (A) posits that the bulk of eukaryotic cellular complexity evolved prior to the acquisition of the α-proteobacterial endosymbiont that became the mitochondrion. This includes the DNA-containing nucleus and endomembrane system, which is used for intracellular cargo transport. In this model, the evolution of a cytoskeleton—an internal scaffolding system—is thought to have been necessary in order for the proto-eukaryotic host to be able to engulf the bacterial progenitor of the mitochondrion by phagocytosis (cell eating). The hydrogen hypothesis (B) posits that the

Fig. 6. Continued. mitochondrion and the eukaryotic cell evolved in concert with one another. An intimate symbiosis based on the complementary metabolisms of two prokaryotes served as the starting point: in the absence of oxygen, a methane-producing archaeal cell (an archaeon) utilized the carbon dioxide (CO_2) and hydrogen (H_2) produced by an α-proteobacterium, which eventually came to reside within the methanogen. As the α-proteobacterial endosymbiont evolved into a dedicated ATP-producing organelle, the archaeal host cell evolved the complex internal features that define the eukaryotic cell. Both evolutionary scenarios involve extensive gene transfer from endosymbiont to host as well as the evolution of a protein import system for targeting the protein products of transferred genes to the newly evolved mitochondrion. The flagellum used by eukaryotic cells for movement has been omitted for simplicity.

digested time and time again along with whatever other cells happened to be around. But these particular α-proteobacteria were 'useful': they metabolized the molecular oxygen before it could cause the host cells harm. The α-proteobacterial endosymbionts persisted within the host cell for progressively longer periods of time without being digested—the host cell was becoming dependent on them. No longer in the mood to digest its source of antidote to the poisonous oxygen, the protoeukaryotic host began experimenting with ways of 'tricking' the endosymbionts into handing over some of their precious ATP; this included adorning their cell surfaces with channels that caused ATP to leak out. The dance had begun. The metabolic interactions between the endosymbiont and protoeukaryote became increasingly intimate.

So too did the genetic and cell biological interactions between the two types of cells. Genes from the genome of the α-proteobacterial endosymbiont found their way into the protoeukaryotic nuclear genome (Fig. 6A). The transfer of such genes, if they were still needed to maintain the endosymbiont's increasingly sheltered existence, would have been lethal (for the endosymbiont donor, not the host). But no matter, provided there were lots of endosymbionts in any given host cell. And when the host started making proteins from these transferred genes and shipping them out to the endosymbionts floating around inside it, no endosymbiont need have a copy of the transferred gene. A dedicated system of protein import evolved; this allowed the endosymbionts to 'outsource' an increasing number of the proteins they needed to live. And with that, the two organisms had become metabolically, genetically, and cell biologically inseparable. The mitochondrion was born. And so was the common ancestor of the complex cell.

Is this how it happened? We don't yet know, but as a working hypothesis for the origin of the eukaryotic cell it has a number of things going for it. It accounts for how the endosymbiont came to reside inside its host (phagocytosis) and why the endosymbiont was so important—it served as an oxygen detoxifier. (This particular aspect of the model stems from the work of Carl Woese and the 'ox-tox' hypothesis of Sweden's Charles Kurland and Siv Andersson.[28]) As we shall see, however, the jury is still out over which actually came first, the eukaryotic cell or the mitochondrion. Let's consider what we know for sure.

We know that the common ancestor of present-day eukaryotes was a remarkably complex cell. A wealth of biochemical and comparative genomic data speak loud and clear to this fact. Researchers have begun to assemble a detailed 'parts list' of the functional systems and 'molecular machines' that form the foundation of eukaryotic cellular complexity. If it were an instruction manual it would be called 'how to make a eukaryote'. We know what the amino acid sequences of many of the proteins involved in sub-cellular protein trafficking and phagocytosis look like, and we have a good idea how they work thanks to investigations using 'model' organisms like yeast. The proteins that replicate DNA and synthesize RNA in eukaryotes are known, and we have a laundry list of the proteins that make up the cytoskeleton. We even know the proteins underlying the molecular mechanisms of sex, the genetic mix and match that takes place when egg and sperm unite. And when we search the genomes of animals, fungi, plants, and protists for the genes that code for these proteins, we find them. The ancestor we share with all other eukaryotes was a complicated organism.[29]

It is conceivable that there are genuine transitional eukaryotes lurking somewhere on the planet waiting to be discovered, organisms akin to Cavalier-Smith's archezoa. And lest we be too quick to dismiss this possibility as wishful thinking, consider that over the past decade researchers have uncovered one new microbial lineage after another—prokaryotes and eukaryotes alike—using a combination of cutting-edge DNA-based techniques and good old-fashioned microscopy. Many of these lineages correspond to known groups of organisms, new twigs on the tree of life; others represent entirely new branches, the microbial equivalent of a new type of primate found hidden in the depths of the jungle.[30] As the astronomer Carl Sagan once said, absence of evidence is not evidence of absence—biologists will continue to explore the natural world and there is no telling what they might find.

All things considered, however, it seems increasingly likely that the organisms needed to bridge the gap between prokaryotes and eukaryotes, the missing links, have long since gone extinct. The period during which the eukaryotic cell evolved was in all probability one of intense experimentation. There may have been one or more evolutionary 'bottlenecks' through which only select lineages passed, by virtue of the fit between their rapidly evolving metabolic and cell biological capabilities and the changing environment. Such bottlenecks may well have been triggered by the rise in oxygen.

Bottlenecks aside, there is another critical point on which we can be sure, and with which we will bring this journey home: all eukaryotes possess mitochondria or organelles derived from them.[31] It has taken researchers the better part of 20 years to come to this realization; it represents a significant shift in perspective. Indeed, the fact that mitochondria appear woven into the very fabric of the eukaryotic cell has inspired fresh thinking about how eukaryotes could have arisen. If the evolution of the mitochondrion was not a consequence of the eukaryotic condition, might it have been the cause?

Membranes and the meaning of mitochondria

At the heart of any evolutionary story we tell about ourselves lies the
origin of the eukaryotic cell.

Maureen O'Malley, 2010[32]

William Martin believes that the key to the evolution of mitochondria—
and the key to eukaryotic life—is hydrogen. Not oxygen, hydrogen.
Martin is a tall, gregarious, impossibly energetic Texan who works
at Heinrich Heine University in Düsseldorf, Germany. He believes
that our infatuation with oxygen stems largely from the fact that
most 'textbook' biochemistry was worked out using rat liver
mitochondria—the experimental system of choice for bioenergetics
studies.[33] In 1998, Martin teamed up with Miklós Müller of the Rockefeller University in New York and put forth the 'hydrogen hypothesis'
for the first eukaryote. The eukaryotic cell, they argued, evolved under
primarily *anaerobic* conditions. The evidence lay in the seemingly
limitless diversity of mitochondrial form and function exhibited by
modern-day eukaryotes.

Müller is a pioneer in the field of protist energy metabolism. In 1973,
while in the lab of Christian de Duve, he characterized an unusual
organelle in the cytoplasm of *Trichomonas*, one of Cavalier-Smith's
archezoans.[34] The organelle was surrounded by two membranes and
lacked cristae. It generated ATP but in a highly unusual fashion—in
the absence of oxygen and with the release of hydrogen gas and
carbon dioxide. Naturally, the organelle became known as the hydrogenosome and Müller became its father.

The evolutionary origin of the hydrogenosome was for many years
an enigma. Back in the 1980s Cavalier-Smith had had no problem
putting *Trichomonas* on his list of primitive archezoa; whatever the
hydrogenosome was, it certainly had nothing to do with mitochondria. Actually it did. During the 1990s, as the biochemical eccentricities
of the organelle came to light, it became clear that the *Trichomonas*
hydrogenosome was a mitochondrion-related organelle. And as the

years passed it also became clear that it wasn't just a one-off freak of nature. Hydrogenosomes and other MROs are, slowly but surely, sprouting up all over the tree of eukaryotes, wherever and whenever eukaryotic cells have evolved in low-oxygen or oxygen-free environments. In the case of *Trichomonas* and related species, that environment is the urogenital tract of humans, cows, and cats. In other protists and fungi it's the rumen of cattle, the guts of cockroaches, and marine sediments the world over. And it is not just unicellular eukaryotes—certain deep-sea animals carry out their entire life cycle in the complete absence of oxygen and appear to harbour hydrogenosome-like organelles.[35]

What is it that links hydrogenosomes to 'normal' mitochondria? Some hydrogenosomes have genomes that speak directly to their evolutionary affinity and biochemical capabilities; others do not. But with much hard work researchers have figured out that hydrogenosomes are themselves surprisingly diverse. The genome-lacking hydrogenosome in *Trichomonas* synthesizes ATP in a simple fashion without the benefit of a proton gradient, while the hydrogenosome of the cockroach inhabitant *Nyctotherus* has a genome and is in essence 'An anaerobic mitochondrion that produces hydrogen'.[36] Further insight has come from investigation of another oxygen abhorring protist, *Blastocystis*, a pathogen that inhabits the human gastrointestinal tract. *Blastocystis* makes its ATP using a genome-containing organelle that possesses biochemical features resembling *both* mitochondria and hydrogenosomes.[37] As strange as it seems, there is a continuum of such organelles in nature: from 'typical' aerobic mitochondria such as our own to the 'anaerobic' mitochondria of certain protists, mollusks, and parasitic worms; through to the ATP-producing hydrogenosomes and hydrogenosome-like organelles; and finally to the stripped down, bare-bones MROs of *Giardia* and *Entamoeba*, which do not make ATP at all.

Martin attaches deep significance to this evolutionary continuum. While many researchers see MROs as 'recent' adaptations to anaerobic environments, Martin believes they are a manifestation of characteristics inherited directly from the very first eukaryote: an ancient, metabolically flexible cell that was capable of growth in the presence *and* absence of oxygen. Under the hydrogen hypothesis,[38] that ancestral cell stemmed from a symbiosis between two prokaryotes (Fig. 6B). The first was an α-proteobacterium that could grow aerobically and, under anaerobic growth, produced hydrogen gas (H_2) and carbon dioxide (CO_2) as waste. The second was an ancestor of a specific type of archaeal cell called a 'methanogen'. The methanogens are well-known inhabitants of anaerobic environments on today's Earth; they use H_2 and CO_2 for fuel and produce the greenhouse gas methane.*

Under low-oxygen conditions the lifestyles of these two prokaryotes would have been metabolically complementary: the methanogenic archaeon fed off the waste of the α-proteobacterium, a situation that was greatly enhanced by close physical contact between the two cells.† The relationship got serious; the α-proteobacterium eventually came to reside *within* the methanogen and, as a result, was now completely dependent on it for energy. Bulk movement of genes from the α-proteobacterial endosymbiont to the methanogenic archaeon provided the host with the biochemical means to feed its new tenant using organic matter obtained from the environment.

The rise in molecular oxygen in the oceans approximately two billion years ago triggered a decline in H_2 concentrations (hydrogen and oxygen together make water). This would have spelled tough times ahead for the methanogen were it not for the fact that, strictly speaking, it was no longer a methanogen—during the course of assimilating the α-proteobacterium it had reinvented itself. It was now free to leave behind a life spent craving hydrogen and avoiding

---

* The methanogens are what give marshlands their distinctive 'eggy' smell. They also live in the digestive tracts of animals where they contribute to the methane released during belching and flatulence.
† This sort of thing happens in nature: within the cytoplasm of certain anaerobic protists, methanogenic archaea make a living by snuggling up to hydrogenosomes so as to maximize H_2 and CO_2 uptake.

oxygen. What was left of the α-proteobacterium became the mitochondrion.

At this point you might be wondering when, during the extensive metabolic and genetic renovations that led to the evolution of this chimeric super-cell, did the *eukaryotic cell* evolve? The answer is *at the same time*. The hydrogen hypothesis posits that the defining intracellular features of present-day eukaryotes (nucleus, cytoskeleton, and so on) evolved during and after the evolution of the mitochondrion (Fig. 6B). No mitochondrion, no eukaryotic cell. That is a bold statement. Before exploring why the evolution of the mitochondrion and the eukaryotic cell might have gone hand in hand, let's step back to consider the evolutionary scenarios we have discussed and reflect on the sobering reality of what it is we are trying to explain: the complex cell evolved only once in the approximately four billion year history of life.

Critics of the hydrogen hypothesis raise the issue of how, in the absence of phagocytosis, the α-proteobacterium could actually have ended up inside the methanogen. But there are in fact examples of bacteria living inside other bacteria in nature—so it *can* happen and it only *had* to happen once. More problematic is the double-barrelled metabolism purported to have existed in the ancestral eukaryotic cell: it is not clear why, and indeed how, this cell would have maintained a functional repertoire of aerobic *and* anaerobic metabolic genes (which cannot be in operation at the same time) such that the common ancestor of all mitochondria was sufficiently endowed that it could have given rise to the diversity of aerobic and hydrogenosome-type metabolisms seen in present-day organisms.

The phagotrophy-based 'ox-tox' hypothesis has issues of its own. The proposed reason why the α-proteobacterial endosymbiont might have been 'useful' to its protoeukaryotic host—oxygen detoxification—seems reasonable, but only to a point. Having earned its keep as an oxygen antidote, the endosymbiont then went on to become an oxygen-consuming machine, a mitochondrion. In full flight, mitochondria are potent generators of damaging reactive oxygen by-products. From the perspective of the host, this would seem to have made the problem of oxygen toxicity *worse*, not better. Critics of the 'ox-tox' hypothesis thus question whether protection from oxygen was in fact the initial glue that bound host and endosymbiont together.[39]

No mitochondrion, no eukaryotic cell; it is an important corollary of the hydrogen hypothesis. In contrast, phagotrophy-based models such as the 'ox-tox' hypothesis place the bulk of eukaryotic cellular innovation *prior to* the origin of the mitochondrion. Is it possible to determine which of these fundamentally different evolutionary paths to complex life is most likely to have occurred? Nick Lane and William Martin believe it is, and the answer lies in the profound significance of the bioenergetic membrane.

For all their extraordinary abundance and metabolic versatility, prokaryotes have remained prokaryotes throughout the history of life on Earth. Why? They are forever destined to remain 'small' and 'simple' because the amount of ATP they can generate is a function of their surface area: the bioenergetic membranes across which they generate the proton gradient required to drive efficient ATP synthesis are their *cell membranes*. If a prokaryotic cell becomes too large, the ATP required to sustain the resulting increase in cell *volume* quickly exceeds its ability to make ATP using the cell *surface*. Eukaryotic cells sidestep the problem of energy demand by having mitochondria: the bigger they get the more mitochondria they can fit inside them and the more ATP they can synthesize.[40] Endosymbiosis changes everything.

The specific metabolic details underlying the hydrogen hypothesis for the first eukaryotic cell may prove to be off the mark. Was hydrogen important or was it oxygen? We may never know for sure. More to the point, it may not matter. By placing the origin of the mitochondrion *before* the evolution of eukaryotic cellular complexity, the hydrogen hypothesis avoids putting the cart before the horse—it provides the only source of cellular energy large enough to have supported the increase in complexity in the first place: the bioenergetic membranes of mitochondria.[41] In essence, the internal complexities of the eukaryotic cell simply could not have evolved without the dramatic increase in ATP that only the mitochondrion could provide. This explains why all present-day eukaryotes have mitochondria or organelles derived from them. It explains why the genomic 'parts list' present in the common ancestor of today's eukaryotic organisms is so large and sophisticated. The primordial host of the mitochondrion was in all likelihood a prokaryotic cell; it would have been impossible to become a complex eukaryote capable of phagocytosis without having a mitochondrion. The single-celled protists would not have evolved; multicellular plants and animals could never have arisen; and we would not be in a position to marvel at how it all took place. Let's now marvel at the other great biological revolution resulting from endosymbiosis: the greening of planet Earth.

7

GREEN EVOLUTION, GREEN REVOLUTION

Viewed from outer space, our planet is a sea of blue and green. Dark blue oceans cover two-thirds of its surface; the rest is green—a bit of desert brown, some icy white at the poles, but mostly rich, luscious, magical green. It wasn't always like this. Time and time again, from fiery beginnings up through the eras and epochs, our emerald planet has reinvented itself. A billion years ago, had we been here to see it, the Earth would have had a very different feel. Not only would the landmasses have been unrecognizable, they would have been various shades of brown and grey. And had we been around 700 million years ago, there would have been little more to see than solid sheets of ice and an equatorial ring of surface water—'snowball Earth' scientists call it. Geologically speaking, the greening of our planet took place decidedly late in the day, within the last 500 million years of planetary history.

But what a revolution it was! Mosses, ferns, and shrubs, grasses and cacti, flowering plants galore, hardwoods and softwoods, on and on. These organisms dominate the world we see. And what strange creatures too, these silent, slow-growing, sun-worshiping giants affixed to the surface of the Earth. They absorb nutrients from the soil and sow their seeds any way they can, on the wind, hidden within tasty fruit, or on the bodies of bees and moths.

For all their abundance and diversity, the land plants are, genetically speaking, but tiny twigs on the evolutionary tree of eukaryotes. Using DNA sequence analysis their history can be traced back to a common ancestor they share with much smaller, simpler aquatic creatures, unicellular green algae. The green algae are themselves the product of a revolution, much less obvious but much more fundamental—the evolution of eukaryotic photosynthesis. At its root, the green revolution is a story about microorganisms and endosymbiosis. From remarkably humble beginnings stemmed the forest of green that now surrounds us.

In this chapter we will trace the threads of photosynthesis in modern-day plants and algae backwards and forwards in time through the rich tapestry of eukaryotic life. As is the case for mitochondria, the light-gathering organelles of photosynthetic eukaryotes—chloroplasts, also known as plastids[1]—appear to have evolved from prokaryotic endosymbionts on a single occasion. The evidence suggests that the host organism was a fully fledged eukaryotic cell into which the cyanobacterial progenitor of the chloroplast was assimilated. As we shall see, the impact of this endosymbiotic event on the biochemistry and cell biology of the primordial alga was profound. Yet the story of eukaryotic photosynthesis does not end there. Indeed, in some ways it is at this point that the story *begins*. Chloroplasts have passed repeatedly from one eukaryotic lineage to another by a process called 'secondary' endosymbiosis, giving rise to some of the most genetically and cell biologically complex organisms known to science—cells inside cells inside other cells whose evolutionary histories are written in their DNA. They are also some of the most ecologically significant organisms on the planet; most of the photosynthesis taking place in the oceans is performed by algae that have 'stolen' their chloroplasts from other eukaryotes. History has shown that chloroplasts have long been in high demand—who wouldn't want to be able to tap into an endless supply of solar energy?

In Chapter 6 we touched on the transformative role cyanobacteria have played in shaping the biogeochemistry of Earth. They are the organisms whose ancestors were responsible for the great oxidation event. Consequently, no discussion of eukaryotic photosynthesis would be complete without considering their biology and evolution. Let's start by giving cyanobacteria the attention they deserve.

Cyanobacteria: life's microbial heroes

Harvard palaeontologist Andrew Knoll considers cyanobacteria to be 'arguably the most important organisms ever to appear on Earth'.[2] The word 'cyanobacteria' stems from the Greek 'blue'; traditionally they were called blue-green algae. It has a nice ring to it, but is misleading from today's perspective because the term 'algae' is now typically used in reference to photosynthetic eukaryotes. And yet it is entirely understandable given the many structural and biochemical similarities they share with the light-harvesting organelles of plants and algae. As we have learned, many of these similarities were apparent to scientists more than 100 years ago—we shall revisit them in this chapter, for they provide some of the strongest evidence for the endosymbiotic origin of chloroplasts presently available.

Cyanobacteria thrive in a variety of marine and freshwater habitats. They are readily found growing anywhere there is water and sunlight—from the open ocean, to steaming hot springs, to lakes and streams. Less well appreciated is the ability of cyanobacteria to inhabit terrestrial environments. They happily grow on rocky surfaces, where they often live symbiotically with fungi in the context of lichens (see Chapter 3).

In terms of outward appearances, cyanobacteria seem to have changed little since they first evolved over two billion years ago. Some exist as solitary single cells while others form rudimentary colonies, clusters of interconnected cells that drift about as a single unit. Of particular ecological significance are cyanobacteria that form filaments, long chains of small, rectangular cells interspersed by larger, rounder ones called 'heterocysts'. It is within the thick-walled heterocysts that atmospheric nitrogen is converted into ammonia, an ecologically important process referred to as nitrogen 'fixation'. (Atmospheric nitrogen must be converted into ammonia and other 'fixed' compounds before it can be used by cells for incorporation into DNA, proteins, and other nitrogen-containing compounds essential for life.) As such, filament-forming cyanobacteria are a rarity among prokaryotes in exhibiting a modest level of 'differentiation'—they are capable of forming specialized cell types with distinct functions. The ability to fix nitrogen is a precious asset among living beings; eukaryotic cells are uniformly incapable of it, and only certain prokaryotes possess the biochemical means with which to carry it out. The cyanobacteria are justifiably well known for their nitrogen-fixing capabilities, but they are well and truly famous for something else entirely: their ability to perform oxygenic photosynthesis.

The term photosynthesis refers to the conversion of solar energy into chemical energy, energy that can be used to drive basic cellular processes. What makes cyanobacteria unique among photosynthetic prokaryotes is the sophistication of their light-gathering apparatus and their use of water as a source of electrons. It is difficult to overstate the importance of electrons in biology—these negatively charged subatomic particles lie at the heart of energy transfer in living systems. Recall that in mitochondria, it is the flow of electrons (acquired from the breakdown of food) along a chain of protein complexes that drives the pumping of protons across the inner mitochondrial membrane. The energy stored in the resulting electrochemical gradient is what fuels ATP synthesis (Fig. 4). Cyanobacteria also use electrons to pump protons and to synthesize ATP. But they also attach electrons to carbon dioxide (CO_2)—this is the critical first step in making food from sunlight. These all-important electrons come from the 'splitting' of water into hydrogen and oxygen.

The process of oxygenic photosynthesis involves two distinct sets of biochemical reactions, named 'light' and 'dark'.* In the light reactions, photons strike the chlorophyll pigments within the 'photosynthetic membranes' of the cyanobacterium. Photon absorption triggers

---

* As their name suggests, the light reactions need light; the dark reactions can take place in the light or the dark.

the splitting of water, release of O_2, and initiates electron flow. The movement of electrons not only results in the synthesis of ATP, but also the production of a powerful electron-carrying compound called NADPH (nicotinamide adenine dinucleotide phosphate).

The dark reactions come next: they take advantage of the ATP and NADPH generated in the light reactions to 'fix' CO_2 and drive the synthesis of sugars. As long as the cell receives light, electrons are transferred to CO_2. When the lights go off, the stores of ATP and NADPH are used up and the synthesis of organic molecules stops.

Oxygenic photosynthesis is a chemical marvel of nature. As sources of electrons go, water is a terrible one—it is the most innocuous of substances, an ultra-stable molecule that does not give up its electrons without a fight. But this is precisely what the first cyanobacteria 'figured out' how to do: use the energy of the sun to pry electrons off of water and force them on to CO_2, converting it into sugar.[3] Oxygenic photosynthesis forms the foundation of the food chain; it is the ultimate source of all our energy. It was the seemingly unlikely utilization of water as an electron donor that led to the rise of oxygen in the atmosphere more than a billion years ago. And the oxygen produced by modern-day photosynthetic organisms is what allows us to 'burn' food in our mitochondria using aerobic respiration. Andrew Knoll is right: cyanobacteria may well be the most important organisms to have ever lived. It is difficult to argue otherwise.

Cyanobacteria and chloroplasts— the photosynthetic ties that bind

Cyanobacteria are responsible for 20–30 per cent of the Earth's photosynthetic output. Chloroplast-bearing eukaryotes account for the bulk of the rest. But they are not the organisms you would necessarily expect. Green though our surroundings may be, the photosynthetic activities of land plants are dwarfed by those of microbes. Approximately half of the planet's total annual photosynthetic output is performed not by terrestrial photosynthetic organisms but by single-celled aquatic algae—phytoplankton. It is an impressive statistic, all the more so given that these microbes account for less than 1 per cent of the total photosynthetic biomass on the planet. No less than 20 per cent of the global photosynthetic fixation of carbon is in fact carried out by a single group of unicellular algae in the oceans, the diatoms. Each year these tiny cells produce as much organic carbon as do all the rainforests on the planet combined. Such is the ecological significance of the hidden microbial majority.

There is no such thing as a typical photosynthetic eukaryote. Beyond the obvious morphological diversity exhibited by trees and plants, considerable variation can be seen from group to group at the level of cells and molecules. An individual plant cell usually contains hundreds of chloroplasts within its cytoplasm, while a single-celled green alga will often harbour only one or two. All chloroplasts use a photon-absorbing pigment called chlorophyll *a*—the main pigment used by cyanobacteria—but chlorophyll *b* is a pigment found only in green algal and land plant chloroplasts. Various other pigment types can be found scattered among diverse algal groups. The precise set of light-harvesting and electron transfer proteins used for photosynthesis also varies from lineage to lineage.

Differences aside, chloroplast-bearing eukaryotes all have one very important thing in common: they photosynthesize in a manner that is, for all intents and purposes, the same as in cyanobacteria. It is this single fact, supported by diverse lines of evidence, that binds together all of the plants and algae on Earth; it is a legacy of the process of oxygenic photosynthesis first 'invented' by cyanobacteria and passed on to eukaryotes by endosymbiosis. Let's take a moment to consider how chloroplasts do what they do.

In his book *Eating the Sun*, Oliver Morton referred to a chloroplast as 'a bag inside a bag'.[4] It is a useful way to think about their internal structure. The outermost 'bag' is simply the pair of membranes that separate the contents of the chloroplast from the rest of the cell. The inner bag is all crumpled up—it is a convoluted set of interconnected membranes that form what are called 'thylakoids'. In plants and green algae the thylakoids look like stacks of coins; in other algae the stacks are wider than they are tall, giving them more of a layered sheet-like appearance. Sliced down the middle, thylakoids resemble railway tracks. Each set of 'tracks' is a cross-section through a pair of closely spaced membranes (Fig. 7).

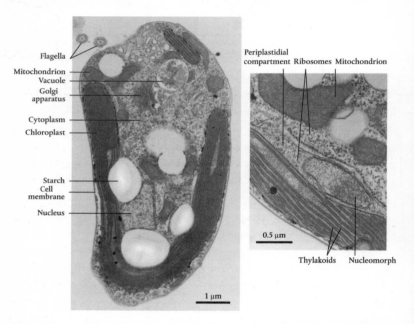

Fig. 7. Transmission electron micrographic images showing the internal structure of a complex algal cell. The organism shown is the cryptophyte alga *Guillardia theta*. A whole cell in cross-section is shown on the left. Note that the long, slender flagella, which these cells use to move through their aqueous environment, appear as circles due to the angle at which they were cut during sample preparation. The image on the right shows a close-up of the chloroplast and associated structures, including the nucleomorph, which is the residual nucleus of an engulfed alga. In this particular type of alga, the periplastidial compartment corresponds to the endosymbiont cytoplasm; it contains ribosomes that synthesize proteins encoded in the nucleomorph genome. The scale bars are in micrometres (μm).

Image credit: Eunsoo Kim.

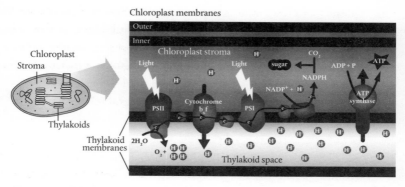

Fig. 8. Simplified schematic diagram showing the process of oxygenic photosynthesis in chloroplasts. Photon absorption by photosystem II (PSII) results in the splitting of water, which produces oxygen gas (O_2) and initiates the flow of electrons (e−) across a series of membrane-embedded protein complexes. Proton (H^+) pumping into the narrow thylakoid space is carried out by the cytochrome b_6f complex, which also serves to pass electrons from PSII to PSI. Photon absorption by PSI leads to the synthesis of NADPH and the addition of electrons to carbon dioxide (CO_2). The resulting sugar participates in a variety of downstream biochemical reactions (not shown). The proton gradient established by the splitting of water and proton pumping activities of the cytochrome b_6f complex drives the synthesis of ATP via the ATP synthase complex, in a manner similar to ATP synthesis in mitochondria.

The thylakoids are typically the most prominent internal feature of a chloroplast, and for good reason: they are where photosynthesis takes place. Embedded within the thylakoid membranes are the so-called photosynthetic reaction centres, the pigment-protein complexes where photons are absorbed and electron transport occurs (Fig. 8). The thylakoid membranes are expanded and crumpled for the same reason that the inner membranes of mitochondria are: to maximize the surface area on which biochemical reactions can take place and across which the all-important proton gradient can be established. Chloroplast thylakoids are derived from the photosynthetic membranes of cyanobacteria, which are also internalized and folded back and forth in order to expand their energetic potential. All things considered, the photosynthetic apparatus functioning in

chloroplasts has a few more bells and whistles than does its counterpart in cyanobacteria, but it does precisely the same job in essentially
the same fashion.

The simple sugars produced by oxygenic photosynthesis in the
chloroplast can be fed into a variety of different biochemical pathways
depending on the needs of the organism. For example, they can be
linked together to form polymers such as starch, which are then
stored away for future use. Alternatively, they can be exported from
the organelle and used immediately as energy to burn. Having focused
our attention on chloroplasts, it is perhaps easy to forget that plant
and algal cells also possess mitochondria, which require organic
molecules to drive ATP synthesis. It is because of their complex
evolutionary history that plants are in the enviable position of being
able to feed their mitochondria without ingesting a thing. They make
their own food from the simplest of ingredients: light, water, and CO_2.
We should be so lucky.

Outsourcing organelles

Ribosomes are the protein-synthesizing 'machines' of the cell. They
comprise dozens of protein and RNA molecules whose structures and
functions are essential for life. Consequently, the gene sequences that
specify these proteins and RNAs change extremely slowly over time,
making them useful for inferring evolutionary history. Recall that in
the 1970s it was the investigation of ribosomal RNA (rRNA) that
allowed researchers to definitively establish the long-suspected endosymbiotic connection between chloroplasts and cyanobacteria. When
compared to one another, the string of chemical letters in chloroplast
rRNAs were found to be remarkably similar to those of cyanobacteria,
somewhat less similar to those of non-photosynthetic bacteria such as
E. coli, and very different from the rRNA sequences encoded by the
nuclear genome.[5] On the basis of this single molecule, the genetic
material in the chloroplast was deemed to share a more recent common ancestor with cyanobacteria than with any other known group

of organisms. An endosymbiotic origin for mitochondria was considered likely based on the results of similar comparisons.[6] An exciting new window on to the evolution of organelles had been opened, but it was only a crack—it took time for the true nature of chloroplasts and mitochondria to reveal themselves. Technology first had to catch up with the ambitions of the researchers.

In 1977, Fred Sanger and colleagues at the University of Cambridge published a new molecular sequencing approach[7] that was to transform modern biology. 'Sanger sequencing' did not require buckets of RNA or protein as starting material—it targeted the DNA itself, the hereditary material in which the genes encoding rRNAs and proteins are stored. DNA is a relatively stable molecule that could easily be purified and sequenced in small quantities. Sanger's method proved so efficient and robust that it wasn't long before scientists were dreaming big, exploring ways in which it could be 'scaled up' and automated.[8] It served as the foundation for the human genome project, the multi-billion dollar initiative to sequence our own genetic material. By the late 1990s, the project was running full tilt, fuelled by sequencing factories around the globe using Sanger's method to churn out millions of base pairs of DNA sequence data per year. The 3.2 billion base-pair sequence of the human nuclear genome was published in 2001.[9]

The Sanger sequencing revolution marked the beginning of the modern era of comparative genomics. Biologists today have the luxury of looking not only at the sequences of individual genes for insight into evolutionary history, they can study an organism's entire *genome*—the complete set of genes it needs to grow and reproduce. Chloroplast and mitochondrial genomes were some of the very first to be sequenced and studied. They proved to be full of surprises.

By the time the first rRNA sequence data from mitochondria and chloroplasts were obtained, molecular biologists had already begun to suspect that their genomes were unusual in several interesting ways. In the late 1960s and early 1970s, researchers had used electron microscopes to investigate organellar genome size and structure: organelles were isolated from cells, broken open, and their DNA molecules spread out on a flat surface for examination, like an entomologist pins down an insect for dissection. Unlike the linear DNA molecules residing in the nucleus, the organellar genomes of various yeasts, plants, and algae were found to be circular molecules.[10] This made sense given that bacterial genomes were also circular. However, organellar genomes were much, much smaller than those of any known bacteria. Precisely why this was the case, and what it meant for the biology of the eukaryotic cell, was the subject of much debate.

How small is small? To put the diminutive size of organellar genomes into perspective, let's consider the bacterium *E. coli*. Its genome is approximately 4.6 million base pairs in length—it has 4.6 million As, Cs, Gs, and Ts arranged in a precise order. Within that 4.6 million base pairs of sequence are ~4,300 genes. These are the complete set of genes that *E. coli* cells need in order to live.

The first organellar genome to be sequenced was the human mitochondrial genome: it was found to be 16.5 *thousand* base pairs in length. It has a grand total of 37 genes, only 13 of which code for protein (the rest are for rRNAs and other small RNAs involved in protein synthesis).[11] With hundreds of mitochondrial genome sequences now in hand from diverse eukaryotic species, it is clear that animal mitochondrial genomes are on the small side of 'normal', but not by much. Even the most gene-rich mitochondrial genome presently known, that of a single-celled protist called *Andalucia*, has only 66 protein-coding genes.[12] The α-proteobacteria, from which mitochondria are believed to have evolved, typically have genomes that encode at least a thousand proteins.

In terms of genetic richness, chloroplasts are somewhat more impressive than mitochondria: algal and plant species typically have between 60 and 200 genes in their chloroplast genomes. They possess genes for making RNA from DNA, genes for building their own ribosomes (rRNA molecules and a number of ribosomal proteins), and genes for photosynthesis, including those specifying various thylakoid proteins. And when possible to discern, these chloroplast genes sing the same evolutionary tune: in phylogenetic trees they branch specifically with their counterparts in the genomes of modern-day cyanobacteria, just as the earliest chloroplast rRNA sequences were shown to do.

Although demonstrably 'cyanobacterial', chloroplast genomes are, like those of mitochondria, mere shadows of their former selves. Free-living cyanobacteria studied to date have between 2,000 and 12,000 genes.[13] It is not yet known to which group of modern-day cyanobacteria chloroplasts are most closely related, but it is clear that chloroplasts contain at most 10 per cent of the genes that were present in the cyanobacterium from which they evolved. Have the genes simply been lost? How many proteins are needed to constitute a fully functioning organelle?

In 1981, before the field of organellar genomics had come of age, the eminent British biochemist John Ellis of the University of Warwick took stock of the processes known to take place within the chloroplasts of plants and algae. It was an impressive list, and getting longer all the time. Ellis speculated that the chloroplast proteins that could be 'visualized' experimentally—200 or so at the time—were in fact 'just the tip of the plastid [chloroplast] protein iceberg'.[14] That same year, an American plant geneticist, Norman Weeden of the University of California at Davis, published a landmark paper in which he addressed the issue of what had happened to the 'missing' chloroplast genes. It was increasingly obvious to Weeden, Ellis, and others that mitochondria and chloroplasts were dependent on far more proteins than their tiny genomes could encode. Weeden proposed two alternative hypotheses:

> One explanation for the small genome of the plastid [chloroplast] is that most of the genes have been lost, their function being replaced by genes already present in the eucaroyte [eukaryote]. In contrast, much of the genome of the endosymbiont could have been transferred to the nucleus during the evolution of the plant cell.[15]

Weeden endorsed the latter explanation, positing that there had been a bulk movement of genetic material from the cyanobacterial progenitor of the chloroplast to the host nucleus. Furthermore, he speculated that the proteins derived from these transferred genes had remained faithful to their roots—despite the fact that they were now nucleus-encoded and synthesized by ribosomes in the host cytoplasm, they still functioned in the organelle from whence they came.

Weeden's instincts proved to be largely correct. As more sequence data accumulated, the nuclear genomes of plants and algae were found to harbour hundreds of genes of cyanobacterial origin. However, the full story proved to be more complicated than Weeden and his contemporaries could have envisioned. Many of the cyanobacterial genes in plant and algal nuclear genomes were indeed found to encode proteins that functioned in the chloroplast, as predicted. The unexpected twist was that not *all* of them did.

In 2002, William Martin and colleagues scoured the nuclear genome sequence of the 'model' land plant *Arabidopsis* for genes of cyanobacterial ancestry.[16] Their goal was to quantify the genetic 'footprint' of the cyanobacterial progenitor of the chloroplast on its host. Approximately 4,500 of the 25,000 genes in the nuclear genome of this particular plant were deemed to be of cyanobacterial ancestry. This was almost 20 per cent of the total number of genes in the organism, an unexpectedly large fraction. What was even more surprising was that less than 50 per cent of these genes seemed to code for proteins that actually functioned in the chloroplast. Proteins of cyanobacterial origin appeared to function all over the *Arabidopsis* cell: in the cytoplasm, in the nucleus, even in the mitochondrion. A similar picture emerged upon investigation of the genomes of other photosynthetic eukaryotes, including single-celled algae. Martin later underscored the potential evolutionary significance of endosymbiont-to-nucleus gene transfer as follows:

> there is no evolutionary 'homing device' that automatically directs the protein product of a transferred gene back to the organelle of its provenance. Instead, the products of genes acquired from endosymbionts can explore all targeting possibilities within the cell.[17]

When it comes to the genetic and biochemical dance that takes place between hosts and endosymbionts early in the evolution of an organelle, the take home message is 'anything goes'.

Jumping genes, shrinking genomes

Molecular sequencing studies have revealed that the genomes of mitochondria and chloroplasts are massively reduced compared to the bacterial genomes from which they evolved. Most of the thousand-plus proteins that function in these organelles are the product of genes residing in the eukaryotic nucleus. By delving into the nuclear genomes of plants and algae, researchers have shown that the cyanobacterial progenitor of the chloroplast served as a source of raw genetic material with which the host cell experimented. The end result of this early 'mix and match' phase of chloroplast evolution is that plant and algal cells are mosaics: many aspects of their biology appear to have been transformed by the prokaryotic endosymbionts that took up residence inside them. In the framework of traditional Darwinian evolution, organisms diverge from one another over time. Endosymbiosis acts to bring evolutionarily distinct lineages *together* in a manner that can lead to the generation of entirely new organisms—one plus one equals one.

With this endpoint in mind, it is now time to consider how, mechanistically speaking, endosymbionts and their host cells actually integrate with one another. How is DNA exchanged? How do the protein products of genes derived from endosymbionts end up functioning in different sub-cellular compartments within the host cell? From a 'nuts and bolts' perspective, there is much to be learned about how endosymbionts become organelles by exploring the biology of present-day life forms.

In 1993 William Martin, Rüdiger Cerff, and colleagues coined the term 'endosymbiotic gene transfer' (EGT) to describe the movement of DNA from endosymbionts to hosts.[18] At its root, the process of EGT stems, somewhat paradoxically, from the fact that cells invest a lot of energy maintaining the integrity of their hereditary material. DNA molecules are long and thin (Fig. 2) and subject to breakage. Cells possess a small army of proteins whose sole job is to detect and repair such damage. It is during the repair process that foreign DNA can get its 'foot in the door': as the ends of broken DNA molecules are being glued back together, the DNA repair machinery can accidentally insert fragments of DNA that happen to be in the neighbourhood, fragments that were not originally in the genome.

In the case of the evolution of photosynthetic eukaryotes, an abundant source of such foreign DNA was the digestion of cyanobacterial endosymbionts. As we will discuss later on in this chapter, there was presumably a phase during the early evolution of algae when cyanobacteria were being ingested by phagotrophic eukaryotes in search of food. These organisms would have been digested shortly after being brought inside the cell. Intracellular breakdown would have made the sugars derived from cyanobacterial photosynthesis available to the host. The digestion of these cyanobacteria would also have released their DNA, most of which would have been quickly broken down into tiny pieces and the individual nucleotides 'recycled'. By chance, however, some of this foreign DNA would have moved into the nucleus before it was completely degraded; here it could be integrated into the nuclear chromosomes by the resident DNA repair machinery. (Note that the nucleus is not completely sealed off from the cytoplasm; it contains numerous pores through which proteins and nucleic acids are constantly coming and going.)

EGT is not just something that scientists have invoked to explain the presence of cyanobacterial genes in the nuclei of modern-day plants and algae—it can be observed in 'real time'. In 2003, research in the laboratories of Jeremy Timmis in Australia and Ralph Bock in Germany provided experimental evidence showing that in certain plants chloroplast DNA jumps into the nuclear genome with astonishing frequency. For example, one in approximately 16,000 tobacco plants harbours a piece of chloroplast DNA in its nuclear genome that its ancestors one generation ago did not have.[19] Sometimes the transferred fragments are only a few dozen nucleotides in length, but they can also be large chunks, tens of thousands of nucleotides long.

In stark contrast, Chris Howe, Saul Purton, and collaborators in the UK showed that in the unicellular green alga *Chlamydomonas*, rates of EGT are apparently so low as to be undetectable in the lab.[20] This variation makes sense when one considers that *Chlamydomonas* cells have a single chloroplast per cell, whereas tobacco cells have hundreds. Such differences in organelle number presumably determine how much chloroplast lysis can take place without killing the organism and, consequently, how much chloroplast-derived DNA can potentially make its way into the nucleus for genetic integration.

EGT involving mitochondrial DNA happens as well. Sequence-based analyses of plant nuclear genomes show that in some cases near-complete mitochondrial genomes have been integrated into the nucleus, and the human nuclear genome is littered with fragments of mitochondrial DNA.[21] Comparative genomic investigations have shown that EGT is an important and widespread phenomenon. In the words of Timmis and colleagues, 'Fragments of organelle DNA are becoming recognized as a normal attribute of nearly all eukaryotic chromosomes'.[22]

All this raises the question of the fate of transferred DNA. If EGT is so common in nature, why, in the fullness of evolutionary time, have organellar genomes not completely disappeared? Why have mito-chondria and chloroplasts outsourced most but not *all* of the genes that underlie their biochemical operations? This issue has puzzled scientists for decades. To properly address it we need to first consider what has to happen in order for a foreign gene to successfully estab-lish itself in a nuclear genome, and what it takes for the protein product of a newly transferred gene to make its way back to its original site of action. Over short evolutionary timescales, the intri-guing answer to the question of what happens to transferred frag-ments of DNA is almost always 'nothing at all'.

The genomes of prokaryotes and the eukaryotic organelles that evolved from them are, in some ways, models of efficiency. They generally pack as many genes as they can into the space they have, and genes involved in similar biochemical processes are grouped together. The nuclear genomes of eukaryotes are typically much less economical—the genes are scattered about in a more or less random fashion, interspersed by large stretches of DNA that do not code for rRNAs or proteins (these are called 'non-coding' regions). Not only are eukaryotic and prokaryotic genomes structured differently, they 'express' their genes in very different ways. Prokaryotes and eukaryotes have different sets of proteins whose job it is to identify a region of DNA as a gene and set about making an RNA copy of it so that a protein can be made. What this means is that even if a prokaryotic or organellar gene finds itself in a nuclear genome there is a decent chance that it will be unrecognizable as such. The transferred gene is often all but invisible to the eukaryotic cellular machinery—its DNA sequence will simply acquire mutations at random and 'drift away' into oblivion.

In considering the fate of transferred genes we must also recognize that just because a fragment of chloroplast or mitochondrial DNA has found its way into the nuclear genome does not necessarily mean that it has been lost from the organelle. As discussed previously, photosynthetic organisms can have dozens to hundreds of chloroplasts per cell, each genetically identical (or very nearly so). Even if a handful of these organelles break open and release their DNA, some of which ends up in the nuclear genome, the organelles that remain intact still possess the stretch of DNA in question. This leads to a situation in which a gene can reside in two different genetic compartments at the same time. Researchers call this promiscuous DNA.

Assuming the copy of the gene in the organellar genome remains functional, the recently transferred nuclear gene can disappear as quickly as it arose without impacting the organism. In the examples of EGT discussed above this is usually what happens. In tobacco, for example, organellar DNA is raining down on the nuclear genome—most of it simply 'runs off' with little or no obvious evolutionary consequence. The same is true in humans: the evidence suggests that the majority of the fragments of mitochondrial DNA in the nuclear genome are inconsequential. Evolutionarily speaking, they are here today, gone tomorrow.

Occasionally, however, it is the *organellar* copy of the gene that is deleted or rendered non-functional by mutation. And it is precisely this situation in which the nuclear version of the gene has the potential to rescue the cell from oblivion. To appreciate how this might happen, we now need to consider the important issue of protein trafficking—how proteins synthesized in one cellular compartment end up functioning in another. How is it that an organelle-derived gene, even if it can be recognized in the host nuclear genome, can produce a protein that can be shipped back to the organelle in which it originally functioned?

A critical step in the early evolution of an organelle is the development of a dedicated protein import apparatus, a set of proteins whose coordinated actions serve to bring in the nucleus-encoded proteins necessary for proper organelle function.[23] Indeed, many molecular biologists consider the presence of protein import machinery to be *the* defining feature of an organelle, the feature that distinguishes it from an endosymbiont.

In modern-day organisms, protein import works as follows. Certain proteins located on the outer surface of the organelle act as gatekeepers: they recognize molecular 'tags' present on the ends of organelle-targeted proteins. It is the presence of a tag—a short stretch of sequence 10–20 amino acids in length—that indicates that a given protein is to be granted access. The gatekeepers then hand off the tag-containing proteins to additional components of the import apparatus whose job is to transport them across the outer organellar membranes, on into the organelle's interior where they can carry out their functions. If the protein product of a transferred gene is to be targeted to an organelle, the gene must acquire a stretch of DNA that specifies the molecular tag that, at the level of protein, is recognized by the import machinery. The import tags used by mitochondria and chloroplasts are biochemically similar yet distinct;

the two organelles have completely different protein import machineries.[24]

Acquiring an organelle import tag is not as difficult as it may seem. Research has shown that there are a variety of molecular mechanisms by which a transferred gene can end up with a tag-specifying sequence. There is a good deal of chance involved. Sometimes a transferred gene simply 'inherits' a tag by virtue of having been inserted into the nuclear genome in just the right place, into a dead or dying gene that already had a tag, for example. In other cases, all it takes is a couple of nucleotide mutations at the front end of the gene to produce a tag-coding region from scratch.

There is also a fair bit of 'slop' in the system: chloroplast-derived genes/proteins can acquire tags that will target them to mitochondria and *vice versa*. And if an organelle-derived gene ends up being expressed in the absence of a tag, its protein has the potential to acquire a function somewhere else, such as in the cytoplasm or the nucleus. Recall the situation described earlier for the plant *Arabidopsis* in which cyanobacterium-derived genes encode proteins that function all over the cell.[25] Endosymbiosis fosters evolutionary experimentation—anything that can happen probably will.

Why do organelles retain genomes?

EGT is a dynamic and ongoing process. It acts as a ratchet, a unidirectional flow of genetic material that has the potential to impact the structure of the nuclear genome in present-day eukaryotes.* As we have learned, in some cases, such as chloroplast-to-nucleus gene transfer in tobacco, EGT is frequent but very little of the transferred DNA seems to 'stick'.[26] Nevertheless, numerous 'recent'—and functional—mitochondrion-to-nucleus gene transfer events have been documented

---

* There are several reasons why DNA can be seen to move from organellar genomes to the nucleus but not in the other direction. The most significant is that while organelle digestion provides a ready source of DNA for transfer into the nucleus, there is no comparable process for the nucleus.

in diverse flowering plant species, including genes specifying ribosomal and electron transfer proteins. Much of this work has been carried out in the laboratory of the American biologist Jeffrey Palmer of Indiana University.[27]

Moving further back in time, we can logically and confidently infer that EGT must have been rampant during the early evolution of mitochondria and chloroplasts, whose genomes today encode at most 10 per cent of the proteins they need to function. Sequence-based analyses tell us that large numbers of genes did indeed move from the bacterial progenitors of these organelles into the host genome.[28] And we have learned how the protein products of transferred genes can acquire the ability to be transported back to their home compartment (or indeed to other sub-cellular compartments). So the question remains: why do mitochondria and chloroplasts retain a genome?

Any organellar gene can, in principle, be relocated to the nucleus. But this does not mean that all of the proteins specified by such genes can be reimported. Biological membranes serve as highly effective water-repelling barriers of macromolecules; they are what biochemists refer to as 'hydrophobic'. Proteins vary considerably in terms of their hydrophobicity—the biochemical characteristics of their amino acid sequences determine how 'comfortable' they are interacting with water molecules. One long-standing hypothesis for why certain proteins remain encoded in organellar DNA is that they are hydrophobic in such a way as to make it difficult for the protein import apparatus to push them all the way across the organelle membranes.[29] Another possibility is that certain organelle-encoded proteins are, for whatever reason, 'toxic' to the cytoplasm of the eukaryote in which they reside.[30] The genes for such proteins can be transferred to the nucleus like any other mitochondrial or chloroplast gene, but when the proteins themselves are synthesized by ribosomes in the cytoplasm they wreak havoc before they can be sequestered and shipped off to the organelle. These genes should, therefore, remain in organellar

DNA; organisms in which one or more of these genes *have* been transferred to the nucleus would fare poorly or would not fare at all.

John Allen, a biochemist at Queen Mary University in London, believes that the genes which remain in mitochondrial and chloroplast genomes are there because it is important for the cell to have them located on site—in the same compartment in which their protein products function. Think back to our earlier discussions of the electron transport chains that lie at the biochemical heart of mitochondria and chloroplasts: they are networks of membrane-anchored protein complexes (each comprised of multiple proteins) that receive chemical inputs from various sources and must act together in a coordinated fashion (Figs. 4 and 8). It is critical that electrons continue to flow through the system; if they do not the chain can 'back up'. With nowhere to go, these electrons can end up attached to oxygen where they give rise to so-called 'reactive oxygen species', by-products of cellular respiration that cause serious damage to DNA, RNA, and proteins.[31]

Given the sheer number of factors involved, it is not surprising that there are numerous ways in which the flow of electrons through bioenergetic membranes can stall. Of specific importance here is a shortage of the electron-shuttling proteins that form critical links in the chain. Many of these proteins are organelle-encoded, particularly in chloroplasts. If the concentration of these key proteins is found to be lacking, the expression of their genes can be increased and fresh protein can be served up in short order. New electron transport chain complexes are assembled, electrons continue to flow, and crisis is averted.

Now consider what would happen if the genes for these key proteins were in the nuclear genome. The organelle sends a 'signal' to the nucleus that proteins X, Y, and Z are needed; the expression of the corresponding nuclear genes is increased and the proteins are synthesized.[32] But how do they get to the specific organelle in which they are needed? The cell does not 'know' which of the potentially hundreds of organelles in the cytoplasm sent the distress call, and so they all receive a shipment of new proteins whether they 'asked' for them or not. To be clear, not all of the electron transport chain proteins are encoded in mitochondrial and chloroplast genomes. Allen's thesis is that the most important ones invariably are—the ones upon which the core membrane-associated bioenergetic complexes are built—and that natural selection favours the persistence of their genes in organellar DNA: 'For safe and efficient energy transduction, genes in organelles are in the right place at the right time.'[33]

Many unicellular algae possess a single chloroplast per cell. Such organisms would therefore not be faced with the problem of having to keep track of which organelle is in need of electron transport chain proteins. Have these organisms relocated the genes in question? In cases where genomic data are available, the answer is 'no'. For example, the *Chlamydomonas* chloroplast genome encodes roughly the same set of proteins that are encoded in the chloroplast genomes of plants, which have numerous organelles per cell. Therefore, the challenges associated with accurate protein trafficking cannot be the sole reason for the retention of organelle genomes. Nevertheless, the issue of efficiency could still be a relevant factor. Synthesizing the key electron transport proteins in the organelle itself, in response to a signal generated in the same sub-cellular compartment, is certainly 'simpler' and would seem to be less prone to time-wasting hang-ups.

Of course, getting rid of the organellar genome altogether is, in its own way, simpler still: no more DNA to replicate and no more energy spent synthesizing and importing dozens of nucleus-encoded proteins for the sole purpose of expressing a small handful of organellar genes. And yet mitochondrial and chloroplast genomes persist. Intriguing support for a specific connection between membrane bioenergetics and the retention of organellar DNA comes from consideration of the very few instances in which organellar genomes *have* disappeared, in the mitochondrion-related organelles (MROs) discussed in Chapter 6. MROs are found in eukaryotes living in low-oxygen or oxygen-free environments. It is significant that in organisms in which the MRO has done away with its genome, the organelle no longer uses an electron transport chain to produce ATP. In the case of the malaria-causing pathogen *Plasmodium*, the mitochondrial genome is almost gone, but not quite: it encodes only three proteins. All three of these proteins are involved in electron transport.

It seems likely that there is no single answer to the question of why organelles retain genomes. But that hasn't stopped scientists from trying to get to the root of the issue—it is, fortunately, an area of evolutionary research in which competing hypotheses make predictions that can be tested in the lab. The precise reasons for genome retention will probably differ between mitochondria and chloroplasts, from gene to gene, and from organism to organism. What we know for sure is that with the exception of a handful of MROs, a small residue of genes remains in mitochondria and chloroplasts, genetic footprints of endosymbiosis still visible after more than a billion years of evolution.

A rainbow of chloroplasts

We are now in a position to step back, shift our focus, and consider the big picture. With the critical pieces of the organelle biology puzzle in hand, we can now address the fundamental issue of how the very first photosynthetic eukaryotes arose and, ultimately, spawned the bulk of the oxygen-producing organisms that now surround us. Researchers have pinpointed the specific branch of the tree of life on which the cyanobacterial thread appears to have first become entwined with eukaryotes. Let's delve into the weird and wonderful world of algae. They are some of the most sophisticated and, it must be said, underappreciated life forms on Earth.

Three eukaryotic lineages are presently recognized as having chloroplasts that stem directly from cyanobacteria. We began this chapter discussing the most prominent among them: land plants and the green algae from which they evolved. Many thousands of green algal species inhabit freshwater and marine environments. They also live on the land where, along with cyanobacteria, they form lichens by living symbiotically with fungi. Among the green algae are some of the smallest eukaryotic cells known, marine algae such as *Ostreococcus*, tiny green balls less than one micrometre in diameter (about the size of an average bacterium). Not surprisingly, *Ostreococcus* and its kin went undetected in the oceans for decades; these so-called 'pico-eukaryotes' are now recognized as abundant and integral players in marine ecosystems. Other species such as *Volvox* are famous for their ability to assemble into beautiful green colonial spheres, each comprising ~50,000 cells. *Volvox* is a close relative of the well-studied green alga *Chlamydomonas*, an organism that spends its life as a single-celled entity. Still other green algae are fully fledged macroscopic life forms; the sea lettuce *Ulva* is a prominent (and edible) example.

The largest of the green algae are not green algae at all, at least not any more. The land plants are genetic offshoots of a specific green algal lineage called the Charales, a relatively simple group of fresh-water pondweeds informally known as 'stoneworts'. Based on fossil evidence and 'molecular clock' analyses, organisms related to present-day Charales appear to have invaded the land approximately 450 million years ago.[34] Such organisms faced both challenges and oppor-tunities. The earliest terrestrial plants would have initially benefited from the higher levels of CO_2 found in the atmosphere than in the water, contributing to the efficiency of their dark reactions of photo-synthesis. But with direct exposure to sunlight, they would have had to develop new ways to protect their cells from the harmful effects of ultraviolet radiation.

The biggest hurdle for land plants to overcome would have been the lack of direct contact with water. Plant cell walls became thicker and enriched with cellulose and lignin, the latter being a complex polymer that provided stiffness to water-filled cells and tissues strug-gling to counteract the effects of gravity (lignin is a significant com-ponent of wood). A complex root system evolved hand in hand with vascular tissue that allowed water to be transported throughout the organism. Symbiotic associations with mycorrhizal fungi living in soil enhanced their ability to take up essential nutrients via their roots.

These were significant and unprecedented evolutionary steps; the land plants transformed the planet in a fashion that was both spectacular and understated. Fuelled solely by photosynthesis, the only way to survive above ground was to spread out and sit still.

Multicellularity evolved independently in another 'primary' lineage of photosynthetic eukaryotes, the red algae. As their name suggests, many (but not all) of these organisms are a deep red in colour, due to their distinctive suite of accessory photosynthetic pigments. More than 5,000 species strong, red algae are a predominantly marine group of organisms that includes seaweeds of considerable economic importance. *Porphyra* is harvested as a wrap for sushi; other species, including *Gracilaria* and *Chondrus* (Irish Moss), serve as invaluable sources of gelling agents such as carrageenan and agar. Carrageenans are used to thicken a wide range of products—everything from ice cream to shampoos—while agar forms the basis of the gelatinous growth medium upon which microorganisms are cultured in labs the world over. A related product, agarose, is an essential laboratory reagent whose natural sieving properties are used in the purification of DNA and RNA.

Somewhat less well known are the pink-coloured coralline red algae that secrete calcium carbonate and form crusts on subsurface rocks; they are important components of coral reefs. Unicellular species of red algae can be found living in hot springs and other extreme environments. For example, *Galdieria* grows in what amounts to battery acid and with direct exposure to arsenic and other toxins.

Members of the red algae appear in several places in the fossil record. Of particular note are 1.2 billion-year-old fossils discovered in the Canadian Arctic by the University of Cambridge palaeontologist Nick Butterfield. Embedded in rock derived from what was once a shallow marine habitat, Butterfield's fossils are very similar in appearance to the modern-day, filamentous, stalk-forming red alga *Bangia*. Naturally, the fossil organism was dubbed *Bangiomorpha*, and the presence of what appear to be fossilized spores suggests that it was capable of sexual reproduction, as is *Bangia* today. If *Bangiomorpha* is what it appears to be, then red algae are the oldest eukaryotic lineage known to exhibit complex multicellularity.[35]

The third and final core lineage of chloroplast-bearing eukaryotes is by far the most rare and enigmatic. Glaucophyte algae are an exclusively freshwater, single-celled group of eukaryotes; a mere dozen or so species are presently known. Under the microscope, the blue-green coloured chloroplasts of glaucophytes look a lot like cyanobacteria, so much so that they were originally assumed to be endosymbionts. Their most unusual feature is the presence of a layer of 'peptidoglycan' between the inner and outer chloroplast membranes. Peptidoglycan is a polymeric, mesh-like component of bacterial cell walls; it is found in cyanobacteria but not in the chloroplasts of red or green algae. As we have learned throughout this book, looks can be deceiving—we now know that glaucophyte chloroplasts are fully integrated sub-cellular organelles, not endosymbionts.

The first photosynthetic eukaryotes

How are the chloroplasts of green, red, and glaucophyte algae related to one another? Are they the product of independent endosymbiotic events involving different cyanobacteria and different eukaryotic hosts? Or was a single type of cyanobacterium engulfed long ago in a common ancestor shared by these three types of algae? These are difficult questions to answer; the biochemical and morphological diversity exhibited within and between these three lineages is considerable. And the fact of the matter is that the diversity of algae in nature extends far beyond the three groups, as we will learn in the next and final section of this chapter. Fortunately there is one very simple way to make sense of all this complexity: biologists recognize the existence of two basic types of photosynthetic organelles in eukaryotes, the differences between them relating to how they evolved and where they reside within the host cell.

'Primary' chloroplasts are the product of primary endosymbiosis—an endosymbiotic event in which a eukaryotic host takes up a prokaryote, in this case a cyanobacterium (Fig. 9A). 'Secondary' endo-symbiosis involves the uptake of a primary chloroplast-bearing alga by a second, unrelated eukaryotic host (Fig. 9B).[36] Primary endosym-biosis has to happen first (if not there would be no algae with primary chloroplasts around to participate in secondary endosymbiosis), and we have already been introduced to the organisms that have them: the green algae (and their land plant descendants), red algae, and glauco-phyte algae. Together these three lineages make up the so-called Archaeplastida (also known as the Kingdom Plantae), one of half a dozen or so main branches on the eukaryotic tree emerging from phylogenetic analyses of genes in nuclear, mitochondrial, and chloro-plast genomes (Fig. 10). In terms of their genetic and cell biological diversity, the Archaeplastida are roughly equivalent to all the animals and fungi on Earth combined.

In addressing the question of primary chloroplast evolution we will focus on insight gleaned from modern biochemistry and comparative genomics. However, it is worth remembering that biologists have been interested in this general issue for many decades. Indeed, much thought had been devoted to the problem of algal diversity prior to widespread acceptance of the endosymbiont hypothesis for eukary-otic organelles. As discussed in Chapter 5, in the 1960s and 1970s Klein and Cronquist had proposed that red algae were the most primitive of all eukaryotes—they descended directly from cyanobacteria and their chloroplasts were of autogenous origin (they evolved from existing components of the cell). The red algae were, Klein and Cronquist argued, the ancestral stock from which all protists, animals, fungi, plants, and algae evolved.[37] At about the same time, other researchers such as Peter Raven argued that red algal chloroplasts were of endo-symbiotic origin, but that they had evolved separately from those found in green algae and plants.[38] The morphology and pigmentation of red and green algal chloroplasts were, quite reasonably, deemed to be too different to have evolved from the same cyanobacterial endo-symbiont. In 1978 Schwartz and Dayhoff came to similar conclu-sions,[39] as did Van Valen and Maiorana in 1980.[40]

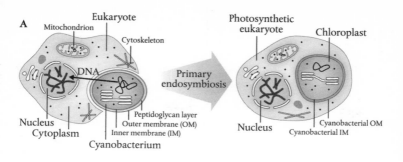

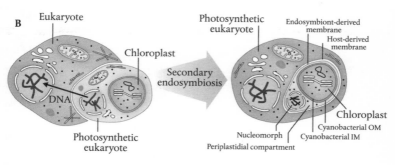

Fig. 9. Endosymbiotic events giving rise to photosynthesis in eukaryotes. (A) Primary endosymbiosis involves a cyanobacterial endosymbiont and a non-photosynthetic host eukaryote. During endosymbiosis DNA moves from the endosymbiont genome to the host nuclear genome, a process referred to as endosymbiotic gene transfer (EGT). The end result of the genetic and biochemical amalgamation of the two organisms is a primary chloroplast-bearing eukaryote. Primary chloroplasts are surrounded by two membranes; both are derived from the cyanobacterial endosymbiont. The peptidoglycan layer present in the cyanobacterial progenitor of the chloroplast was lost in red and green algae, but persists in glaucophyte algae. (B) Secondary endosymbiosis involves the endosymbiotic uptake of a primary chloroplast-bearing alga by an unrelated non-photosynthetic eukaryote. A second wave of EGT occurs, this time from the endosymbiont nucleus to the secondary host nucleus. In certain algae, such as cryptophytes (Fig. 7), the endosymbiont nucleus persists in a miniaturized form called a 'nucleomorph'. The nucleomorph resides in the so-called periplastidial compartment, which is derived from the cytoplasm of the engulfed eukaryotic alga. In other secondary chloroplast-bearing algae, such as diatoms and *Euglena*, nucleus-to-nucleus gene transfer has gone to completion and the nucleomorph has been lost. The mitochondrion of the secondary endosymbiont has also been eliminated. Compared to the primary chloroplasts of green, red, and glaucophyte algae, secondarily evolved chloroplasts have one or more additional membranes. The outermost (fourth) membrane is of host cell origin, acquired during the process of phagocytosis ('cell eating'). When present, the third membrane (from the inside) is thought to correspond to the cell membrane of the ingested eukaryote (i.e. the secondary endosymbiont).

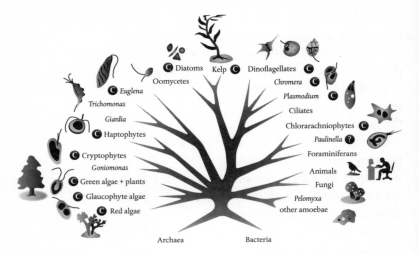

Fig. 10. Schematic evolutionary tree depicting the relationships between some of the eukaryotic lineages discussed throughout this book. Chloroplasts (C) arose by primary endosymbiosis in a common ancestor shared by green algae (and land plants), red algae, and glaucophyte algae. Multiple secondary endosymbiotic events resulted in the spread of chloroplasts across the eukaryotic tree. Numerous branches have been omitted for clarity. There is still much uncertainty about the deepest divisions of the tree; the position of the so-called 'root' of the tree of eukaryotes has yet to be resolved. Readers wishing to learn more are encouraged to explore the primary literature cited in the main text.

Molecular data changed everything. Evolutionary analyses of chloroplast-encoded rRNA and protein sequences have shown not only that green, red, and glaucophyte chloroplasts all descend from cyanobacteria, but that they are specifically related to one another.[41] That is, the three groups branch as each other's closest neighbours in phylogenetic trees, close to, but clearly distinct from, the cyanobacteria. As noted earlier, chloroplast genomes typically have between 60 and 200 protein-coding genes. The red algal and glaucophyte organelles are the most gene-rich; satisfyingly, the 'missing' genes in green algae and plants can often be found in their nuclear genomes, clear-cut examples of 'recent' EGT. The size, structure, and gene content of chloroplast genomes in green, red, and glaucophyte algae suggest that they are the product of the same endosymbiotic event.

Further support for the notion of a single origin of primary chloroplasts comes from consideration of the protein import apparatus. With nuclear and chloroplast genome sequences now in hand, researchers have assembled 'parts lists' for the import machineries in the three algal groups to search for clues as to how they evolved. Not surprisingly, many of the individual membrane-associated import proteins come from cyanobacteria, the inference being that they were inherited early on as genes were being moved from the endosymbiont into the host nucleus and as the precise mechanisms of protein import were being established. A core set of these cyanobacterial-derived import proteins are shared between red, green, and glaucophyte algae. Of particular interest is the existence of two key import proteins (tic 110 and toc 34) found in all three lineages that do *not* have obvious counterparts in cyanobacteria.[42] These proteins appear to be host cell-derived evolutionary innovations—proteins that were recruited from the host into new functions in protein import for the newly evolved organelle. The presence of these alga-specific proteins in all three of the primary chloroplast-bearing groups makes it unlikely that their import apparatuses evolved independent of one another. All things considered, a single origin of primary chloroplasts is the most likely scenario.[43]

How did it happen? Primary chloroplasts are truly ancient; they are thought to have evolved at least 1.5 billion years ago.[44] Nevertheless, a reasonably detailed picture of the initial interactions between host and endosymbiont has emerged from the synthesis of various lines of evidence. The host eukaryote—a common ancestor of green, red, and glaucophyte algae—was a complex, unicellular phagotrophic cell. It was capable of ingesting other microbes for food; cyanobacteria would presumably have been among those taken up and digested. The sugars produced by the photosynthetic activities of these cyanobacteria would have been of considerable benefit to the host, but at this stage they could only be accessed if the ingested cells were broken down.

Insight into the early fate of the cyanobacterial endosymbiont that gave rise to the chloroplast comes from consideration of the mechanics of phagotrophy (cell eating). In present-day eukaryotic cells, ingested food particles end up being surrounded by a host-derived membrane, a so-called phagocytic membrane that forms during the internalization of the outer cell envelope (picture a small soap bubble forming inside a larger bubble by being drawn in and pinched off). It is within these membrane-bound compartments that food is digested. In the case of chloroplasts, it appears that the cyanobacterial endosymbiont somehow 'escaped' the confines of its phagocytic compartment. A significant feature of the primary chloroplasts of green, red, and glaucophyte algae is that they are surrounded by two membranes (Fig. 9), both of which, research suggests, are derived from the cyanobacterial endosymbiont—there is no third membrane as would be expected if a cyanobacterium with two outer membranes were engulfed by a eukaryote. What this means is that the host cell would have lost the ability to digest the cyanobacteria within it.

None of this would have happened overnight. There was very likely a period of time during which cyanobacteria were being taken up and digested, over and over and over again. Gradually, presumably quite fortuitously, a population of cyanobacteria established themselves in the host cell cytoplasm. And this is where they would stay. The earliest steps in chloroplast evolution are often couched in terms of 'predator–prey' interactions, but it can just as easily be seen from the perspective of colonization: the eukaryotic cytoplasm was simply a new habitat for the cyanobacterium to conquer, one that was relatively stable and offered shelter from predation.

Symbiosis is all about give and take. In the case of the early evolution of chloroplasts, an important question is how the host cell managed to gain access to the fruits of cyanobacterial photosynthesis—sugar—without digesting its endosymbionts. Comparative genomic and biochemical investigations have provided insight into how this critical metabolic connection might have been established. The membranes of modern-day chloroplasts possess a diverse suite of transporter proteins that mediate the movement of chemical intermediates into and out of the organelle. Debashish Bhattacharya, Andreas Weber, and colleagues have shown that the original chloroplast sugar transporters appear to have evolved from pre-existing components of the eukaryotic host, in particular endomembrane-associated transporter proteins.[45] Various other types of metabolite transporters known to function in chloroplasts are clearly of cyanobacterial ancestry.

As first noted by Weeden,[46] whom we met earlier in our discussions of EGT, many of the host–endosymbiont connections established during the course of early algal evolution had nothing to do with photosynthesis. For example, metabolic pathways for the synthesis of essential amino acids such as tryptophan, as well as the production of fatty acids and isoprenoids, are today localized in the chloroplasts of green, red, and glaucophyte algae. The overall picture emerging is one of mosaicism: both host- and endosymbiont-derived proteins played a role in the forging of permanent biochemical links between the two organisms. The development of a protein import apparatus, itself a mix of protein subunits derived from both endosymbiotic partners, would have greatly increased the frequency with which cyanobacterial genes established themselves in the host nucleus, further reducing the degree of autonomy exhibited by the cyanobacterium. By this point it was no longer an endosymbiont—it was an organelle.

And it was from a single ancestral lineage of photosynthetic eukaryotes that the green, red, and glaucophyte algae emerged. The layer of peptidoglycan that surrounds the glaucophyte chloroplasts was lost in reds and greens; the light-harvesting machineries of all three groups were subject to evolutionary tinkering as the organisms adapted to the demands of photosynthesis in strange new environments. Land plants evolved from within the green line; seaweeds of the intertidal zone evolved from unicellular red algae; and the glaucophyte algae, low in numbers but rich in evolutionary significance, specialized in freshwater habitats. And as the primary lines of algae continued to diversify, some of them became permanent fixtures inside other eukaryotic cells via secondary endosymbiosis. During the billion-plus-year timescale of algal evolution, chloroplasts have hopped from branch to branch, resulting in the emergence of new forms of life.

Chloroplasts on the move—endosymbiosis drives algal diversification

The single-celled alga *Euglena* is a fascinating organism. Bright green and relatively large for a microbe, it shape-shifts through its environment, pencil-thin one moment, balloon-shaped the next. *Euglena* is a favourite among biology teachers: take a scoop of pond water, bring it back to the lab, and chances are your students will find it. *Euglena* was probably among the first microorganisms ever seen by a human being. 'Little animals' fitting its description were observed in the 1600s by the Dutchman Antonie van Leeuwenhoek and his home-made microscopes that we learned about in Chapter 1. And yet biologists have long struggled to classify *Euglena*. Its colour suggested it was a green alga, but in other ways it was most unlike green algae and indeed any other type of photosynthetic eukaryote. Simply put, *Euglena* was an algal misfit.

In 1977, McGill University Professor Sarah ('Sally') Gibbs was asked to review a scientific manuscript describing the process of cell division in *Euglena*. Gibbs was as qualified a reviewer as anyone. She was a phycologist—she studied algae. For more than a decade Gibbs had been using electron microscopes to characterize the sub-cellular structure of photosynthetic eukaryotes. She knew where their chloroplasts sat in the cell relative to the nucleus, she knew the shape of their chloroplast thylakoids, and she knew how many membranes enveloped them.

Gibbs and her fellow researchers knew that *Euglena* was different. But the *Euglena* chloroplast was known to contain chlorophyll *b*—the signature light-harvesting pigment of green algae and plants—and so that is what it was: a green alga. Among other things, what struck Gibbs as peculiar was that *Euglena* had three membranes around its chloroplast, not two as in the case of green algae. As she read through the manuscript, she struggled to reconcile her knowledge of chloroplast diversity with *Euglena*-style cell biology. And then it hit her:

> I suddenly realized that *Euglena* wasn't related to green algae at all. It just ate them for supper. Probably lots of times, for many euglenoids are phagotrophic, but one time at least a green alga escaped being digested and became established as a permanent endosymbiont...The moment I thought of it, I was ecstatic. I knew it had to be true. It explained too many things not to be true.[47]

What had got Gibbs so excited was the possibility of secondary endosymbiosis, the idea that a eukaryotic cell could acquire photosynthesis by assimilating a primary chloroplast-containing eukaryote.[48] From earlier work Gibbs knew that there were at least half a dozen different classes of algae with not three but *four* membranes around their chloroplasts. She speculated that in at least some of these cases the outermost membrane was not of cyanobacterial origin (as it is in primary chloroplasts), but was a phagocytic membrane derived from the host cell. She pondered the extent to which secondary endosymbiosis might have played a role in the diversification of algae.

As Gibbs dug deep into the literature on endosymbiosis she discovered that she was not the first person to have entertained the notion of eukaryotes evolving inside eukaryotes. In 1974, Max Taylor (whose Serial Endosymbiosis Theory was discussed in Chapter 5) had also mused about a possible secondary origin of chloroplasts in *Euglena*.[49] In the mid-1970s, Dennis Greenwood and colleagues of Imperial College, London, published a pair of obscure abstracts describing the internal structure of a group of red-pigmented algae called cryptophytes (Fig. 7). Of particular interest was Greenwood's description of a tiny membrane-bound structure nestled up against the cryptophyte chloroplast: he called it a 'nucleomorph'. In the most cautious of terms, Greenwood suggested that the nucleomorph was the nucleus of a eukaryotic endosymbiont and that it might have a genome.[50] In 1985, Gibbs and Martha Ludwig showed that Greenwood was correct: the cryptophyte nucleomorph did indeed contain DNA, although at the time its evolutionary origin could not be determined. A few years later, Susan Douglas, Michael Gray, and colleagues at Dalhousie University in Canada filled in the missing blank by demonstrating the existence of two distinct eukaryote-type rRNA genes in cryptophyte cells, one corresponding to the nucleomorph, the other to the host nucleus. Today nucleomorphs are hailed as the 'smoking guns' of secondary endosymbiosis (Fig. 9B), the tiny remains of an endosymbiotically derived nucleus that is going, going, but not quite gone. In the case of *Euglena* and a variety of other algae, the nucleomorph *has* gone.

Various lines of evidence show that the secondary endosymbiont that gave rise to the cryptophyte nucleomorph and chloroplast was a red alga. A collaboration between research teams led by Douglas, Uwe Maier in Germany, and Tom Cavalier-Smith in the UK culminated in the 2001 publication in *Nature* of the first nucleomorph genome sequence, that of the cryptophyte *Guillardia*.[51] The genome was shown to be extraordinarily reduced: it has ~500 genes, less than 10% of the number found in the smallest nuclear genomes of free-living algae. In 1984 the phycologists David Hibberd and Richard Norris discovered a chloroplast-associated nucleomorph in another algal group, the chlorarachniophytes (the term chlorarachniophyte roughly translates to 'green spider'). In this case the nucleomorph and chloroplast proved to be of green algal ancestry. Geoff McFadden, Paul Gilson, and colleagues in Australia sequenced the nucleomorph genome of the 'model' chlorarachniophyte *Bigelowiella*: it was found to contain even fewer genes than does the red algal-derived cryptophyte nucleomorph genome.[52]

The mechanics of secondary endosymbiosis are reasonably well understood. Recall that cyanobacterium-to-nucleus EGT was an important factor in the evolution of primary chloroplasts (Fig. 9A). In the context of secondary endosymbiosis, a second wave of EGT takes place: endosymbiont-derived genes residing in the primary algal nucleus jump across to the secondary host nucleus (Fig. 9B). In many algae, such as *Euglena*, diatoms, and kelp, the gene transfer process has gone to completion and the nucleomorph has completely disappeared. All that remains in such organisms is the chloroplast, a tiny residue of endosymbiont-derived cytoplasm, and one or two additional membranes surrounding the two cyanobacterium-derived membranes (that is, three or four chloroplast membranes in total).

A long-standing debate in the field of algal cell evolution is the question of why nucleomorphs persist. If nucleomorphs have vanished in some secondary chloroplast-bearing algae, why have they not disappeared in all of them? Insight has come from exploration of nuclear genome sequences from the cryptophyte *Guillardia* and the chlorarachniophyte *Bigelowiella*. While instances of very recent mitochondrion-to-nucleus gene transfer were found in both genomes, researchers failed to uncover a single case of chloroplast-to-host-nucleus or nucleomorph-to-host-nucleus gene transfer.[53] This suggests that the rates of EGT between the two nuclei harboured within cryptophyte and chlorarachniophyte cells are extremely low. It appears that the nucleomorphs in these two algal lineages are, for the moment at least, 'frozen'—their genomes contain a small number of essential genes, the majority of which encode proteins dedicated to the synthesis of an even smaller number of proteins targeted to the chloroplast.

One important aspect of the evolution of secondary chloroplast-bearing algae is the manner in which they 'solve' the problem of chloroplast protein import. Again, the solution is one of mix and match biochemistry. Secondary host-nucleus-encoded proteins destined for the chloroplast have two molecular 'tags'. The first tag grants them passage through the outermost chloroplast membrane using protein machinery derived from the eukaryotic host; the second tag moves them into the chloroplast proper using the same mechanism (and the same protein import machinery) that exists in the primary chloroplasts of green, red, and glaucophyte algae. Unlike the cyanobacterial endosymbionts that gave rise to the very first chloro-plasts, secondary endosymbionts come with the genes for chloroplast protein import already in place—they solved that problem long ago. All that needs to happen is for the secondary host cell to contribute some of its pre-existing protein trafficking machinery to the task. Significantly, examination of chloroplast protein import pathways in independently evolved secondary chloroplast-bearing algae reveals that this exercise in biochemical repurposing has happened roughly the same way each time.

The primary endosymbiotic origin of eukaryotic photosynthesis appears to have been a singularity in evolution. The process of sec-ondary endosymbiosis served to spread chloroplasts from the photo-synthetic roots to the furthest tips of the eukaryotic tree (Fig. 10).[54] The consequences were significant. Some of the algae that acquired chloroplasts in this manner contributed to a transformation of Earth's biosphere. The haptophyte algae and diatoms, for example, are among the most abundant oxygen-producing phototrophs in the world's oceans. Other organisms, such as dinoflagellates, seem to do a bit of everything: they are bloom-formers, toxin-producers, predators, para-sites, and, in the context of coral reefs, symbiotic partners with invertebrate animals.[55] Still others have abandoned photosynthesis entirely. Apicomplexans such as the malaria parasite *Plasmodium* evolved from algae—their non-photosynthetic, red algal-derived chloroplasts are so interwoven with the biochemistry of the host cell that they cannot be lost. The photosynthetic ties that bind each of these lineages together are ancient; recognition of the global sig-nificance of endosymbiosis comes from consideration of their recently evolved idiosyncrasies.

8

BACK TO THE FUTURE

I call this experiment 'replaying life's tape.' You press the rewind button and, making sure you thoroughly erase everything that actually happened, go back to any time and place in the past—say, to the seas of the Burgess Shale. Then let the tape run again and see if the repetition looks at all like the original.

Stephen J. Gould, 1989[1]

It is in our nature to want to understand our place in the world, to know where we came from and where we are going. We crave answers to the biggest of the big questions. What is life? What makes us human? Are we alone in the universe? In this book we have explored four billion years of Earth history, focusing on two apparent singularities whose evolutionary significance cannot be overstated. First and foremost, the complex eukaryotic cell evolved from simpler prokaryotic cells once during the entire history of life. The precise details of the prokaryote-to-eukaryote transition remain shrouded in mystery; it nevertheless seems increasingly likely that it was coincident with—and a direct consequence of—the endosymbiotic origin of mitochondria and the power of their bioenergetic membranes. Second, a single primary endosymbiosis gave rise to chloroplasts and oxygenic photosynthesis in eukaryotes. What followed was a chain of events that led to a transformation of ocean, land, and atmosphere.

What are we to make of these singularities? If, as envisioned by Gould, life's tape were rewound and replayed, what are the chances that it would unfold in the same way? The historical nature of evolutionary biology makes this question extremely difficult to address. Occasionally, however, nature provides us with the means to obtain fleeting glimpses of an answer, to assess the relative contributions of chance and necessity in the evolution of life. In this penultimate chapter we will learn how studies of newly established host–endosymbiont relationships in nature have breathed new life into the age-old debate over how mitochondria and chloroplasts evolved, and whether they could ever have arisen the same way twice.

Organelle redux

Biologists use the term 'organelle' in a variety of ways. In principle it can be applied to any part of a cell that carries out a specific function or set of functions; this includes everything from ribosomes to nuclei. We have used the term to refer specifically to membrane-bound compartments that evolved by endosymbiosis—mitochondria and chloroplasts. Thus far, however, we have sidestepped the question of what an endosymbiotically derived organelle *actually is*. What is it, precisely, that allows one to conclude that the transition from endo-symbiont to organelle has occurred? On this point there is both clarity and uncertainty, as highlighted by recent genomic investigations of an enigmatic amoeba named *Paulinella chromatophora*. This organism pos-sesses photosynthetic inclusions derived from cyanobacteria; as we shall see, whether the term 'chloroplast' should be applied to them is a matter of taste.

Paulinella has a long history in scientific circles. It was discovered on Christmas Eve 1894 by the German biologist Robert Lauterborn (1869–1952) who found it while sifting through the polluted sediments of the river Rhine (he named the organism in honour of his stepmother Pauline). Lauterborn was interested in the biology of freshwater microbes, in particular those that feed on dead organic matter—he would spend years investigating the ecology of the Rhine as part of a long-term study on the impact of industrial and domestic pollution. On this particular occasion he found a testate amoeba (an amoeba with a shell) unlike anything that had been seen before. What was most unusual was the presence of blue-green pigmented bodies inside

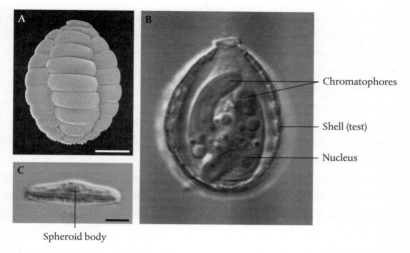

Fig. 11. (A) Transmission electron micrographic image of *Paulinella chromato-phora*, a testate (shell-covered) amoeba with photosynthetic organelles derived from cyanobacteria. The scale bar represents 3 micrometres (μm). (B) A light micrographic image of *Paulinella* showing internal structures. Each cell contains two sausage-shaped chromatophores. Analysis of the chromatophore genome shows that it is the product of a primary endosymbiosis that was independent of that which gave rise to the classical chloroplasts of green algae, red algae, and glaucophyte algae. (C) Light micrographic image of *Rhopalodia* sp., a diatom (a unicellular alga) with a cyanobacterial endosymbiont called a 'spheroid body'. The scale bar represents 10 μm.

Image credits: Takuro Nakayama (A and C) and Eva Nowack (B).

it, two per cell (Fig. 11A and B). Lauterborn was struck by the similar-ities between the pigmented bodies and cyanobacteria; in his official description of *Paulinella* he referred to them as 'chromatophore-like structures'.[2] (This was ten years prior to the first explicit proposal for an endosymbiotic origin of chloroplasts by the Russian Constantin Mereschkowsky.)

For the next 100 years *Paulinella* featured only occasionally in the literature. In 1990, Ingo Reize, a student in the laboratory of Michael Melkonian of the University of Cologne, Germany, succeeded in establishing a laboratory culture of the organism, greatly enhancing the prospects for rigorous study.[3] By this time the chromatophores of *Paulinella* had been shown to possess a cyanobacterium-like peptido-glycan wall (as found in the chloroplasts of glaucophyte algae), and to divide synchronously with their host. Indeed, they appeared incapable of living on their own if released from the amoeba cytoplasm. The amoeba itself was similarly hooked; *Paulinella* depends on the photo-synthetic activities of its chromatophores for survival.

Melkonian's team published the first molecular data from *Paulinella* in 1995. Analysis of nucleus-encoded ribosomal RNA (rRNA) sequences revealed that the organism was not a traditional alga—it was not related to the red, green, and glaucophyte algae that harbour the primary chloroplasts discussed in the previous chapter. Instead, *Paulinella* was found to belong to the same branch of the eukaryotic tree as the chlorarachniophytes, the 'green spider'-like unicellular organisms with secondary chloroplasts derived from green algae (Fig. 10).[4] Was the *Paulinella* chromatophore also of secondary endo-symbiotic origin? It was not. In fact it was not specifically related to chloroplasts at all. Phylogenetic analysis of endosymbiont-derived rRNA gene sequences showed the chromatophore to be related to a specific subgroup of cyanobacteria called *Synechococcus*, highly distinct from classical chloroplasts. The chromatophore was hailed as 'a plas-tid [chloroplast] in the making', a second instance of a primary endosymbiotic origin of photosynthesis in eukaryotes.[5]

Not everyone was convinced. Ursula Theissen and William Martin cautioned that while endosymbioses are common in nature, endo-symbiotic events in which prokaryotes are actually converted into organelles are extraordinarily rare—the evolution of mitochondria and chloroplasts being the only two known examples.[6] They pointed to a 1985 paper by Tom Cavalier-Smith and John Lee in which a distinction between endosymbionts and organelles was proposed: an endosymbiont, they argued, is not truly an organelle until it has evolved the ability to import the protein products of genes that have been transferred to the host nuclear genome.[7] In the case of *Paulinella*, the relationship between the cyanobacterium-derived chromatophores and the host amoeba was clearly intimate. But it was far from clear where the chromatophore fitted along the spectrum between ingested food particle and bona fide organelle.

A significant advance in our understanding of the evolutionary history of the *Paulinella* chromatophore came from its complete genome sequence, published by Eva Nowack, Michael Melkonian and Gernot Glöckner in 2008.[8] The genome was shown to be approximately one million base pairs in size (with 867 protein-coding genes), much larger than any known chloroplast genome, but only a third the size of the genomes of free-living *Synechococcus* species, which have well over 3,000 genes. The chromatophore genome was found to possess a large suite of genes dedicated to photosynthesis, but genes for numerous metabolic pathways were absent, including those for the synthesis of certain essential amino acids. Most interesting was the paucity of genes for membrane-associated proteins involved in the transport of small molecules. Collectively, these observations fuelled speculation that the host amoeba was actively contributing to the biochemical operations of its chromatophores. In isolation, however, the chromatophore genome could not answer the question of whether it was an endosymbiont or an organelle.

The critical pieces of the chromatophore puzzle came from investigation of the *Paulinella* nuclear genome. Evidence for chromatophore-to-host-nucleus gene transfer was first presented in 2009 by Takuro Nakayama and Ken-ichiro Ishida of Tsukuba University in Japan,[9] and a subsequent large-scale study by Eva Nowack and colleagues revealed the presence of more than 30 genes of chromatophore origin in the amoeba nucleus, most of which encode small proteins involved in photosynthesis.[10] Together with Arthur Grossman at the Carnegie Institution for Science, Stanford, Nowack provided experimental evidence showing that at least three of these nucleus-encoded proteins are indeed targeted to the *Paulinella* chromatophore.[11]

The endosymbiotic event that gave rise to the chromatophore appears to have taken place within the last 100 million years;[12] it is clearly in the early stages of integration with its host organism, but the chromatophore is a photosynthetic organelle by any reasonable definition. Some researchers are content to call it a chloroplast (or plastid). Others feel that a different name should be used in order to make it clear that it has a different evolutionary history than the primary chloroplasts of red, green, and glaucophyte algae.

There is much hinging on the biochemical nuts and bolts of protein import in *Paulinella*, most of which remains to be elucidated. The experiments of Nowack and Grossman show that at least some of the nucleus-encoded proteins make their way to the chromatophore via the host cell's membrane-based cargo trafficking system. Intriguingly, bioinformatic investigations carried out in the laboratory of Andrzej Bodyl of the University of Wrocław in Poland suggest that some of the cyanobacterial proteins that more than a billion years ago evolved to function in chloroplast protein import may also have been recruited for the job in the *Paulinella* chromatophore.[13] If this can be verified experimentally it would represent a striking example of parallel evolution, an instance in which the same solution was arrived at to solve the problem of evolving a protein import apparatus for a prokaryotic endosymbiont.

These apparent instances of biochemical recycling are reminiscent of what was described in Chapter 7 in the case of secondarily evolved chloroplasts; let's revisit them briefly here. Unlike primary chloroplasts, which stem directly from cyanobacteria, the chloroplasts of algae such as *Euglena*, cryptophytes, diatoms, and certain dinoflagellates are the product of eukaryote–eukaryote endosymbioses (Fig. 9). The chloroplast protein import pathway in these organisms is a hybrid of both host- (that is, eukaryotic) and endosymbiont-derived machineries—the process is necessarily more complex because their chloroplasts require a two-tiered system to get nucleus-encoded proteins across the extra membranes surrounding them. The number of secondary endosymbioses that have occurred during eukaryotic evolution is uncertain, but an absolute minimum of three and perhaps as many as eight or more evolutionary mergers must be invoked to account for all of the chloroplast-bearing eukaryotes shown in

Fig. 10.[14] In each and every one of these mergers, the same basic host-derived cargo transport machinery was co-opted to act in chloroplast protein import. In the case of the chromatophore in *Paulinella*, the experimental data of Nowack and Grossman suggest that the host cell's cargo transport system was utilized as well. Bodył and colleagues suggest that *Paulinella* may in fact use several distinct mechanisms to shuttle proteins to the outer surface of its chromatophore and across its membranes to the interior of the organelle.[15]

The extent to which the *Paulinella* chromatophore provides a window on the evolution of classical primary chloroplasts is unclear. However, there is no doubt that it represents an important system with which to study evolution in action. Not only are there free-living relatives of the cyanobacterium-derived chromatophore in nature, the amoeba itself has a close cousin with an intriguing lifestyle. Unlike *Paulinella chromatophora*, *Paulinella ovalis* does *not* have chromatophores; instead it makes a living by ingesting and digesting cyanobacteria of the genus *Synechococcus*,[16] precisely the same group of organisms from which the chromatophore evolved. With all of the relevant players in hand, research has begun in earnest to explore the earliest steps in organelle evolution using a system that, in certain respects at least, appears to mirror the initial events leading to the rise of photosynthesis in eukaryotes.

Another fascinating endosymbiosis between eukaryotes and cyanobacteria involves a freshwater diatom by the name of *Rhopalodia gibba*. Inside the diatom lives an endosymbiotic cyanobacterium that is closely related to free-living cyanobacteria belonging to the genus *Cyanothece*—the endosymbiont is referred to as the 'spheroid body' because of its shape (Fig. 11C). *Cyanothece* species are notable for their ability to fix atmospheric nitrogen, that is, to convert nitrogen gas into ammonia and related compounds. Although less well studied than *Paulinella*, the *Rhopalodia*-spheroid body endosymbiosis is clearly a stable relationship in which the two partners are dependent upon one another. In this case the host organism was already photosynthetic at the time the endosymbiosis was established: diatoms harbour chloroplasts derived from the secondary endosymbiotic uptake of a red alga. Not surprisingly then, the spheroid body no longer carries out photosynthesis; what it appears to do is fix nitrogen like free-living *Cyanothece* species.[17] This provides a ready explanation for why a photosynthetic endosymbiont would be of value to the diatom (eukaryotic cells do not carry out nitrogen fixation but are neverthe-less dependent on nitrogen-containing molecules for life, DNA and proteins being the most significant examples).

The endosymbiosis between *Rhopalodia* and *Cyanothece* was estab-lished even more recently than the event that gave rise to the chro-matophores of *Paulinella*, perhaps as recent as 12 million years ago.[18] Uwe Maier and colleagues in Marburg, Germany, have begun to characterize the spheroid body genome; they estimate it to be approximately 2.6 million base pairs in size, large for an endosymbi-ont but nevertheless about half the size of the genome of its closest free-living *Cyanothece* relative.[19] Interestingly, genes for light-harvesting proteins can still be found in the spheroid body genome, inactive and accumulating mutations—a legacy of its past photosynthetic capacity. Whether or not the spheroid body should be considered an endosym-biont or an organelle in the strictest sense will depend on what studies of the nuclear genome of *Rhopalodia* uncover. Regardless, it clearly carries out a highly specialized function for its diatom host.

Endosymbionts galore—anything goes in the game of life

Paulinella and *Rhopalodia* are but two of many known examples of endosymbiosis in nature. I have highlighted them together in order to emphasize the differences in the biology of their respective host organisms at the time of endosymbiosis, and how this relates to the probable selective value of establishing a permanent relationship with an intracellular bacterium. In the case of *Paulinella*, the host amoeba was non-photosynthetic; natural selection would presumably have favoured amoebae that, for whatever reason, delayed the digestion of the photosynthetic *Synechococcus* species they were eating. The establishment of mechanisms for exchanging metabolites between the host and endosymbiont would have been an important factor in converting the engulfed cyanobacterium into a chromatophore. In contrast, the diatom *Rhopalodia* was already a photosynthetic organism. While it is not clear how *Cyanothece* ended up inside the host eukaryote, its nitrogen-fixing abilities would have been a precious acquisition, enhancing its ability to grow and reproduce.

Apart from mitochondria and chloroplasts, *Paulinella* is presently the only case of endosymbiosis involving uptake of a prokaryote in which the term organelle comfortably applies. However, one cannot help but wonder whether this is simply due to lack of study. There may well be other instances of newly evolved organelles that will emerge upon close inspection. Indeed, the more we learn about the diversity of recently established endosymbioses, the less clear the distinction between endosymbiont and organelle becomes.

At the extreme low end of the spectrum is the phenomenon of kleptoplastidy—'plastid stealing'. This occurs when an organism harvests chloroplasts from an alga and keeps them for a relatively short period of time. Kleptoplasts are not organelles, nor are they technically even endosymbionts: they are not stably inherited and so must be reacquired periodically by ingesting algal prey. The most striking—and controversial—example of kleptoplastidy is in sea slugs such as *Elysia chlorotica*. Juvenile slugs of this species suck the chloroplasts out of the cells of a multicellular alga called *Vaucheria*. The slugs are very picky eaters; they will graze only on this particular organism (different species have different food preferences). Having done so, they sequester the chloroplasts within the cells lining their digestive tract and from that point on live like a plant: bright green with chloroplast pigment, back fanned out like a leaf, seemingly fuelled with nothing more than light and air for up to ten months. Kleptoplasty is not optional for *Elysia*: it is essential for the development of the animal, something it must do to become an adult and complete its lifecycle.

Despite decades of research, no one has figured out quite how *Elysia* is able to do this. In essence, the mystery revolves around how the sequestered chloroplasts remain photosynthetically active without support. The critical point here is that when the slug feeds on the alga, the algal nucleus is discarded. Therefore, the chloroplast cannot be restocked with the hundreds of essential chloroplast proteins that are encoded in the algal nuclear genome. For its part, the *Vaucheria* chloroplast genome encodes only 139 proteins, a genetic endowment similar to most other algae. We can thus rule out the possibility that the *Vaucheria* chloroplast is unusually well equipped to sustain itself while moonlighting in the body of the slug.

To explain the long-term maintenance of kleptoplasts in *Elysia* and related sea slugs, researchers once thought that genes for essential chloroplast-targeted proteins had been transferred from the algal nucleus to the slug nucleus during the course of feeding. The evidence currently in hand suggests that this is not the case.[20] What is clear is that as much as they appear to be photosynthetic animals, the slugs must reacquire their chloroplasts with each generation. How much of their nutrition in the wild actually comes from the photosynthetic activities of their kleptoplasts remains to be seen, and seems to vary considerably from species to species. (The amount of time the klepto- plasts are retained by the slugs is in fact quite variable; in some lineages they are digested more or less immediately, in other cases they persist for a few weeks. In being able to keep its kleptoplasts for months at a time *Elysia chlorotica* is unusual.)

Kleptoplastidy is also observed in unicellular eukaryotes such as dinoflagellates. These organisms are remarkable for their ability to capture and recapture chloroplasts by endosymbiosis; they are often referred to as the cellular equivalent of 'Russian nesting dolls'.[21] Only about half of known dinoflagellate species are considered photosyn- thetic. Those that are exhibit the full spectrum of chloroplast types, from transient kleptoplasts to fully integrated organelles serviced by a full suite of nuclear genes for chloroplast-targeted proteins. Many dinoflagellate species are somewhere in between, in possession of a stably inherited chloroplast and yet still actively ingesting algae as prey. Their penchant for phagocytosis may explain why dinoflagellates so readily acquire chloroplasts from other algae and convert them into their own organelles. Their nuclear genomes are often populated with genes for chloroplast-targeted proteins that appear to have come from a variety of different algal sources, genetic evidence in support of their promiscuous ways. All things considered, the dinoflagellates embody what Larkum, Lockhart, and Howe refer to as the 'shopping bag' model of chloroplast evolution, whereby repeated, transient endosymbioses can result in the acquisition of foreign genetic material, which in turn paves the way for the eventual establishment of a stable host–endosymbiont relationship. The organelle that results is a chimaera:

> Your shopping might all be in a bag that came from an identifiable store, and some of the contents of the bag might have come from the same place. But some came from elsewhere, and you cannot ascribe a single origin to all your shopping.[22]

A recurring theme in endosymbiosis and organelle evolution is genome reduction. This refers to the shrinkage that occurs when an endosymbiont loses genes that are no longer essential for intracellular life and, in some cases, transfers genes to its host. In this vein we shall now consider one final type of endosymbiosis common in nature, the so-called nutritional symbioses that exist between sap-sucking insects and bacteria. These are of considerable interest to biologists because of the extreme levels of genome reduction seen in the endosymbionts, and the resulting mix and match biochemistry that melds the insects and their bacteria together into a single functioning unit.

Many insects belonging to the order Hemiptera, such as aphids and mealybugs, have evolved to feed on plant sap. The German biologist Paul Buchner (1886–1978) first noted the association between the highly specialized diet of such insects and the presence of bacterial endosymbionts (some of the best studied of these bacteria belong to the genus *Buchnera*, named in his honour). Within the insect the endosymbionts are housed in a specialized organ called a bacteriome, which is itself comprised of special cells called bacteriocytes—it is within the bacteriocytes that the endosymbionts live, bathed in nutrients and sheltered from the outside world. In return, the bacteria synthesize essential amino acids, which the insect is utterly dependent upon because of the poor nutritional value of sap. This arrangement has evolved multiple times during the course of insect evolution and, once established, is highly stable: in some cases the endosymbionts have been vertically inherited through the maternal line for hundreds of millions of years.[23] Comparative genomics has served as a powerful tool for elucidating the biology and evolution of these insects and their microbial companions.

When it was sequenced in 1995, the genome of the human bacterial pathogen *Mycoplasma genitalium* was considered to be near the lower limit of what was possible for a self-sustaining organism. It was 580 kilobase-pairs in size and found to encode only 470 proteins. With any fewer than about 300 proteins, researchers predicted, core cellular processes such as protein synthesis and DNA replication would be irrevocably compromised. Over the past decade investigations of the genetic make-up of insect nutritional symbionts have torn the bottom out of the once-popular 'minimal genome' concept.[24]

Unlike *Mycoplasma*, which can be grown on its own in the lab, the nutritional symbionts are highly adapted to life within the insect bacteriome. It was thus not surprising to see that their genomes were reduced. However, as more and more genomes are sequenced the record for the fewest number of genes in a cellular organism continues to fall. As of July 2013, the reigning champion in microbial minimalism is a bacterium called *Nasuia deltocephalinicola*, one of two symbionts that live and work together in the bacteriocytes of a leafhopper insect. Gordon Bennett and Nancy Moran of Yale University showed that the *Nasuia* genome is a mere 112 kilobase-pairs in size and encodes 137 proteins. Its partner, *Sulcia muelleri*, has 190 protein-coding genes in its 190 kilobase-pair genome.[25] What *Nasuia* and *Sulcia* are able to do together is produce the ten essential amino acids their host insect cannot get from its diet—*Sulcia* makes eight of them, *Nasuia* the other two. The burning question is: with so few genes how are the endosymbionts themselves able to survive? There is presently no simple answer, but the general scope of the problem and possible solutions have emerged from consideration of the *Sulcia* and *Nasuia* genomes together with those of various other nutritional symbionts.

From the proteins that remain encoded in bacterial endosymbiont genomes less than approximately 300 kilobase-pairs in size, it is often unclear how these organisms manage to carry out critical tasks such as building functional ribosomes, repairing their DNA, and fuelling their stripped down metabolism. In many cases the genes for proteins involved in proton pumping and electron transfer are missing or damaged, and so ATP synthesis would seem to be severely compromised. One emerging possibility is that the insect produces sugar transport proteins that allow its own energy-containing compounds to be passed to the endosymbionts within its bacteriocytes. It is also possible that key proteins involved in ATP synthesis that are missing in one symbiont are somehow obtained from the other, or that ATP is shunted from the insect to the bacteria directly.[26]

What about endosymbiotic gene transfer (EGT)? Could the protein products of endosymbiont-derived genes in the insect nucleus compensate for those lost from the endosymbiont? As in the case of sea slug kleptoplastidy discussed previously, the answer seems to be a qualified 'no'. The nuclear genomes of two sap-feeding insects have been completely sequenced and examples of functional endosymbiont-to-nucleus gene transfer have not been identified. In some cases, however, the host insect has acquired genes from *other* bacteria—that is, bacteria that are not presently living within its bacteriome. Research carried out by John McCutcheon, Takema Fukatsu, Carol von Dohlen, and colleagues have shown that the nutritional symbiosis happening within the mealybug *Planococcus citri* is maintained by a combination of genes from no fewer than six evolutionarily distinct organisms: the insect, two endosymbionts living in its bacteriocytes (one bacterium lives inside the other!), and three other bacteria that contributed genes to the nuclear genome of the insect.[27]

When it comes to describing such complex interspecies metabolic networks biologists are presently at a loss for words. Strictly speaking, the bacteria living inside sap-sucking insects are not organelles. And yet at less than 200 kilobase-pairs in size, the genomes of a growing number of them are *smaller* than those of many of the classical organelles discussed throughout this book, such as the chloroplasts of red algae. Indeed, there is a sense in which the biochemical crosstalk that takes place between different organisms in the context of a nutritional symbiosis is no different than that which occurs between the discrete compartments of a eukaryotic cell—cytoplasm, mitochondrion, and, when present, chloroplast. With our growing knowledge of endosymbiosis, the boundary between endosymbionts and organelles has become blurred.

Chance and necessity

We began this chapter pondering Gould's 'tape of life' experiment and the singular origins of mitochondria and chloroplasts. Through the lens of comparative genomics, we have searched the diversity of host–endosymbiont relationships in nature for clues as to how these hallmark eukaryotic organelles might have arisen. We have found evidence for the existence of defined evolutionary pathways along which endosymbionts evolve and organelles arise from them. Genomes shrink as cells adapt to life inside other cells; genes move from endosymbionts to hosts but not in the other direction. Membrane-associated proteins are recruited to the task of intercellular metabolic exchange, the raison d'être of endosymbiosis. Mechanisms for protein import evolve, often by the functional reassignment of proteins carrying out similar jobs elsewhere in the cell. And organellar genomes persist—a small but critical residue of genes remains locked in mitochondrial and chloroplast DNA because of the biochemical properties of the proteins they encode. These are the universal themes of organelle evolution, of how cells evolve within other cells.

Evolutionary mergers involving eukaryotic hosts and eukaryotic endosymbionts have occurred on numerous occasions during the diversification of photosynthetic life. The bioengineering problems 'solved' by one eukaryote (such as protein import) are passed on to the new host organism. The new host builds upon what it inherits using what it already has; evolution does not look forwards but what it *does* do is work with what is presently at hand. Sometimes nature reinvents the wheel. Often times it does not. Studies of secondary and tertiary endosymbiosis show this to be the case.

And yet there is chance at every turn of the path. Endosymbionts are common but organelles are rare. Primary chloroplasts evolved once from a cyanobacterial endosymbiont approximately 1.5 billion years ago, a testament to the complexities associated with evolving a eukaryotic organelle from a prokaryotic cell. A photosynthetic organelle of primary endosymbiotic origin has nevertheless evolved in *Paulinella* within the last 200 million years, providing an opportunity to explore the early stages of this fundamental process.

What of the evolution of mitochondria? It was indeed a singularity, a chance encounter between two very different types of cells approximately two billion years ago, one ending up inside the other. The biochemical characteristics of the partner organisms are as yet unclear; we have no good present-day model systems with which to investigate how it might have happened. It is nevertheless clear that genetic exchange between the two partners was not only extensive it was essential, a necessary precondition for the evolution of the internalized bioenergetic membranes that fuel complex life. From such membranes came the energy to expand the host's genome, the freedom to expand its suite of genes, providing the raw material to experiment with what might be, to become a complex, compartmentalized cell. It only happened once. It is entirely possible that it would not happen if life's four billion year tape were replayed.

What remains is for us to ponder the existence of two anciently evolved, fundamental eukaryotic organelles—mitochondria and chloroplasts—and a continuum of more recently evolved, intimate associations between all manner of organisms: eukaryotes and prokaryotes; eukaryotes and eukaryotes; predators and prey; grazers and phototrophs; microbes and multicellular animals and plants. It was the endosymbiotic origin of mitochondria and the emergence of the complex eukaryotic cell that made it possible.

9

EPILOGUE

In late February 1975 a landmark scientific gathering was held at the Asilomar Conference Center on California's Monterey Peninsula. A joining of two Spanish words, Asilomar means 'refuge by the sea'; nestled among weathered pines and shrubs, its rustic buildings flank the sand dunes and rocky shores of the Pacific Ocean. Researchers have been flocking to this peaceful setting for decades. It's the perfect place to escape the day-to-day grind of academic life and rejuvenate one's passion for science. On this particular occasion, however, the 146 delegates had a lot on their minds. They were an eclectic mix—mostly biochemists and molecular biologists but also lawyers, government officials, and journalists. They had come to Asilomar to discuss what to do about recombinant DNA technology.

A few years prior to the conference, Paul Berg, biochemist and Professor at Stanford University, had pulled the plug on an experiment in which he and his graduate student Janet Mertz were to infect *E. coli* cells with a virus. It was a virus that had never existed before, an artificially created genetic hybrid whose genome was part *E. coli* virus and part SV40—a monkey virus known to cause tumours in rodents. Scientists knew that viruses mediated the transfer of genes between bacteria in nature. Berg wondered whether a virus-mediated gene transfer system also existed in mammalian cells. The hybrid virus had been created to explore this possibility. An interesting question was whether it could also infect *E. coli*.

Upon hearing about Berg and Mertz's research plans, several colleagues raised concerns about the possible consequences. *E. coli* is a natural inhabitant of the human intestinal tract; what if *E. coli* infected with the chimeric virus somehow escaped from the lab and caused a cancer epidemic? The chances of that happening, Berg and Mertz felt, were extremely remote. Even if the hybrid virus *did* escape there was no evidence to suggest that it presented a risk to human health. Nevertheless, the experiment was set aside until the potential dangers were better understood.

Berg and Mertz were not the only scientists making chimeric DNA molecules. Stanley Cohen at Stanford and Herbert Boyer at the University of California, San Francisco, had been experimenting with how to insert foreign DNA into *E. coli* plasmids—circular, naturally occurring DNA molecules in bacterial cells. Their announcement in 1973 that a ribosomal RNA gene from a frog had been 'cloned' and propagated in *E. coli* stirred significant debate within the scientific community and, eventually, the general public.[1] It was clear that genetic material from essentially any organism on the planet could be spliced into an *E. coli* plasmid in the lab. The possibilities for fundamental and applied research were endless: *E. coli* could be used as a 'factory' for the production of life-saving drugs; a myriad of practical issues in agriculture and industry could be addressed; our ability to explore the basic processes of life would be transformed. But there was also cause for concern. In 1974 Berg and a dozen of his colleagues called for a moratorium on certain types of genetic engineering until the issue of public safety could be addressed.[2]

The main goals of 'the Asilomar Conference', as it became known,* were to assess the potential biohazards of such recombinant DNA research and to decide whether the self-imposed moratorium should be permanent. It was by all accounts an intense and stressful event. There were far more questions than answers. In the end, the decision was made to lift the moratorium. Genetic engineering of the sort pioneered by Berg, Mertz, Cohen, and Boyer would resume under a strict set of laboratory guidelines drafted by the Asilomar attendees

---

* There were actually two 'Asilomar' conferences on this topic in the 1970s, the first a low-profile affair held in January 1973.

and subsequently approved by governing agencies around the globe—the greater the perceived hazard, the more rigorous the safety practices to be implemented. The biotechnology industry was born; the foundation for the genomics revolution was laid (cloning quickly became an integral component of Sanger sequencing). Life science research was transformed.

Forty years on, advances in biotechnology continue at breakneck speed. While the basic public health and safety issues associated with genetic engineering are now considered negligible, the legal, ethical, and environmental implications remain actively debated. And so they should. But what is often forgotten is that the raw ingredients for recombinant DNA research are natural. They are harvested from the microbial biosphere. The DNA cutting protein *Eco*RI comes from *E. coli*: it functions naturally as a cellular defence agent against foreign genetic material such as DNA injected by viruses. Hundreds of such proteins have been identified and are used routinely in diverse research applications. Proteins used to glue DNA fragments back together in the lab are part of the cell's natural DNA repair machinery. A protein called *Taq* polymerase is used to 'amplify' DNA so that it can be studied more easily. *Taq* comes from *Thermus aquaticus*, a bacterium discovered in 1965 living at 70°C in a hot spring in Yellowstone National Park. In the cell, *Taq* polymerase is a DNA replication enzyme. In a test tube, *Taq* will make millions of identical copies of any DNA molecule you give it, no matter what the source. *Taq* has become a cornerstone of molecular biological research; it is used in everything from medical diagnostics to crime scene investigations, from genealogical studies to environmental monitoring.

All of this would be impossible were it not for the common ancestry of life. And what we have learned from the application of recombinant DNA technology to the study of genomes is that genetic exchange is *part of life*. The impact of horizontal gene transfer on the genetic make-up of microorganisms is sufficiently large that researchers now debate the very existence of prokaryotic 'species'. In 2002, a three-way comparison of sequenced *E. coli* genomes—from the benign laboratory strain K12, from a strain infecting the urinary tract, and from the infamous food-borne pathogen O157:H7—painted a remarkable picture. While the ribosomal RNA gene sequences of the three strains are identical, only 40 per cent of the genes were found to be present in all three.[3] With each new sequenced genome the set of genes common to all E. coli gets smaller and smaller. Genes appear to be coming and going all the time; to say that prokaryotic genomes are dynamic hardly seems adequate. These are not merely academic curiosities: the emergence of multidrug-resistant bacteria by gene exchange concerns us all. It is evolution in action.

We are evolution in action—16.5 thousand base pairs and 13 protein-coding genes is all that remains of our mitochondrial genome. And yet its bacterial ancestry can be traced back two billion years to a time before there were animals, plants or fungi, or even single-celled prot-ists as we presently recognize them. As in the vast majority of other eukaryotes, our mitochondria lie at the energetic heart of our cells—their impact on our lives as human beings is significant. Their funda-mental properties impact how we age, how our cells commit suicide. Mutations in mitochondrial DNA are associated with a wide range of serious diseases. So too are the random fragments of mitochondrial DNA that frequently migrate to our nuclear genomes, re-enacting a critical aspect of how our mitochondria first evolved, wreaking havoc if and when they land near or within important nuclear genes.[4]

We are natural chimaeras to the core of our being. As animals we belong to a short twig on the tree of life; as eukaryotes we are part of an ancient genetic and biochemical web, intimately connected to the microbial biosphere. It is a fact of life, a fact of our lives, so easy to forget but so important to remember.

GLOSSARY

Archaea (archaebacteria) A group of single-celled organisms, morphologically similar to bacteria (eubacteria) but evolutionarily distinct from them. The archaea are one of the three domains of life.

ATP Adenosine triphosphate, the 'chemical currency' of all cells, found in prokaryotes and eukaryotes alike. Cells produce ATP using a variety of different biochemical processes, including respiration (e.g. in the mitochondrion of eukaryotes) and fermentation.

Bacteria One of two groups of prokaryotic organisms, the other being archaea (archaebacteria). The bacteria are one of the three domains of life.

Chloroplast The sub-cellular compartment within the cells of plants and algae in which photosynthesis takes place. They are surrounded by two or more membranes and contain their own genome. Chloroplasts evolved by endosymbiosis from cyanobacteria. The term chloroplast is often used interchangeably with plastid.

Cyanobacteria An important group of marine and freshwater bacteria that carry out oxygenic photosynthesis. They can be single-celled or colonial.

Endosymbiosis A form of symbiosis in which two cells live together in nature, one inside the other. The organism living inside is referred to as an endosymbiont.

Eukaryotes Life forms comprised of one or more complex cells, each containing various organelles including a nucleus, cytoskeleton, mitochondrion, and (in plants and algae) a chloroplast.

Genome All the genetic material possessed by an organism or organelle derived by endosymbiosis. This includes all the genes that encode proteins and structural RNA molecules such as ribosomal RNA, as well as non-coding regions of DNA. Viruses also have genomes, which can be made of DNA or RNA.

Mitochondrion The structure in eukaryotic cells in which respiration and ATP synthesis takes place. Mitochondria are surrounded by two membranes and possess a genome. Like chloroplasts, mitochondria evolved from bacteria by endosymbiosis.

Nucleus A membrane-bound structure in eukaryotic cells in which most of the cellular DNA is stored.

Organelle A discrete structure inside a cell in which specific cellular functions are localized. Examples are the nucleus, mitochondrion, and chloroplast of eukaryotic cells.

Phylogenetics The use of molecular sequence data (DNA, RNA, or amino acid sequences) to infer the evolutionary history of genes present in particular organisms, often with the goal of inferring how the organisms themselves are related to one another.

Photosynthesis The biochemical process by which solar energy is converted to chemical energy. It occurs in algae and plants as well as a variety of bacteria. Oxygenic photosynthesis (as carried out by cyanobacteria and algal/plant chloroplasts) produces molecular oxygen, anoxygenic photosynthesis does not.

Plastid See chloroplast.

Prokaryotes Organisms whose cellular structure is defined mainly by the absence of a DNA-containing nucleus and other organelles. Bacteria and archaea are prokaryotes.

Respiration A set of metabolic processes that organisms use to convert the biochemical energy present in nutrients into adenosine triphosphate (ATP), the 'chemical currency' of living cells.

Ribosome A large sub-cellular complex that serves as the site of protein synthesis in all cells. Comprised of numerous proteins and several RNA molecules, the ribosome strings together individual amino acids (in the order specified by the genes) to produce protein molecules. Ribosomes are essential for life: they are found in eukaryotes, prokaryotes, mitochondria, and chloroplasts.

RNA Ribonucleic acid, a linear molecule similar in structure to DNA. In living cells RNA molecules exist in a variety of different forms: some are structural, such as the ribosomal RNAs (rRNAs) that are part of the ribosome and participate in protein synthesis. Messenger RNA molecules serve as the bridge between the genetic information stored in DNA and protein: they are copied from DNA and 'read' by the ribosome to produce protein molecules.

ENDNOTES

1. Life as We Don't Know It

1. Letter 39 from Antonie van Leeuwenhoek to the Royal Society of London, dated 17 September 1683.

2. Revolutions in Biology

1. Schrödinger, E. 1944. *What is Life?* Cambridge University Press.
2. Timoféeff-Ressovsky, N.W., Zimmer, K.G., and Delbrück, M. 1935. Über die Natur der Genmutation und der Genstruktur. *Nachrichten von der Gesellschaft der Wissenschaften zu Göttingen, Mathematisch-physikalische Klasse, Fachgruppe VI: Biologie* 1, 189–245. For English translation and extensive discussion see: Sloan, P.R. and Fogel, B., eds. 2011. *Creating a Physical Biology: The Three-Man Paper and Early Molecular Biology.* University of Chicago Press. See also Perutz, M.F. 1987. Physics and the Riddle of life. *Nature* 326, 555–8.
3. Stent, G.S. 1966. Waiting for the Paradox, in *Phage and the Origins of Molecular Biology*, Cairns, J., Stent, G.S., and Watson, J.D., eds. Cold Spring Harbor Laboratory Press.
4. Symonds, a geneticist at the University of Sussex, worked with both Schrödinger and Delbrück in the late 1940s to early 1950s. Schrödinger told Symonds 'Either you learn some quantum theory or you do something quite different. I would suggest you try biology'. Symonds, N. 1986. What is Life?: Schrödinger's Influence on Biology. *The Quarterly Review of Biology* 61, 221–6.
5. De Duve, C. 2005. *Singularities (Landmarks on the Pathways of Life).* Cambridge University Press.
6. According to molecular biology pioneer Frank Stahl, the 'Phage Church' as they were sometimes known, was 'led by the Trinity of Delbrück, [Salvador] Luria, and Hershey. Delbrück's status as founder and his ex cathedra manner made him the pope, of course, and Luria was the hard-working, socially sensitive priest-confessor. And Al (Hershey) was the saint.' Stahl, F.W. 2001. Alfred Day Hershey: Biographical Memoirs. *National Academy of Sciences (U.S.).* 80, 142–59.

7. Hershey, A. and Chase, M. 1952. Independent Functions of Viral Protein and Nucleic Acid in Growth of Bacteriophage. *Journal of General Physiology* 1, 39–56.

8. Avery, O.T., MacLeod, C.M., and McCarty, M. 1944. Studies on the Chemical Nature of the Substance Inducing Transformation of Pneumococcal Types: Induction of Transformation by a Desoxyribonucleic Acid Fraction Isolated From Pneumococcus Type III. *Journal of Experimental Medicine* 79, 137–58.

9. Watson, J.D. and Crick, F.H.C. 1953. A Structure for Deoxyribose Nucleic Acid. *Nature.* 171, 737–8.

10. 'Chargaff's rules' state that the proportion of A residues and T residues in DNA is the same, as is C and G. This information, apparently conveyed to Watson and Crick by Chargaff himself ('they impressed me by their supreme ignorance'), was said to have helped them solve the structure of DNA, see Judson, H.F. 1979. *The Eighth Day of Creation: Makers of the Revolution in Biology.* Simon and Schuster. This has been debated. Chargaff had not expounded upon the possible significance of his own data and was not awarded a Nobel Prize. He spent much of his career railing against the failings of molecular biology, which he referred to as the 'practice of biochemistry without a license'.

11. Watson, J.D. 1968. *The Double Helix.* Penguin Books USA Inc.

12. Watson and Crick. 1953. A Structure for Deoxyribose Nucleic Acid.

13. Judson, *The Eighth Day of Creation.*

14. Meselson, M. and Stahl, F.W. 1958. The Replication of DNA in *Escherichia coli. Proceedings of the National Academy of Sciences USA* 44, 671–82.

15. Nirenberg, M.W. and Matthaei, H.J. 1961. The Dependence of Cell-Free Protein Synthesis in *E. coli* Upon Naturally Occurring or Synthetic Polyribonucleotides. *Proceedings of the National Academy of Sciences USA* 47, 1588–602.

16. Crick, F.H.C. 1958. On Protein Synthesis. *Symposia of the Society for Experimental Biology* 12, 138–63.

17. In certain instances RNA can serve as a template for DNA synthesis, one of several examples of 'non-standard' information transfers that have been discovered. However, DNA-to-RNA-to-protein is far and away the dominant mode of information flow in molecular biology. With respect to the term 'dogma', Crick acknowledged that he was wrong to use it in this context: 'I just didn't *know* what dogma *meant*.' See Judson, *The Eighth Day of Creation.*

18. Certain codons have, on rare occasions, been 'reassigned' to a different amino acid, such as in the genomes of certain mitochondria. All things considered, however, the nature of the genetic code speaks strongly to the fact that all life on Earth is related.

19. Crick, On Protein Synthesis.

20. See the following historical perspective and references therein: Stretton, A.O.W. 2002. The First Sequence: Fred Sanger and Insulin. *Genetics* 162, 527–32.

21. Zuckerkandl, E. and Pauling, L. 1965. Molecules as Documents of Evolutionary History. *Journal of Theoretical Biology* 8, 357–66. See also the following overview and references therein: Morgan, G.J. 1998. Emile Zuckerkandl, Linus Pauling, and the Molecular Evolutionary Clock, 1959–1965. *Journal of the History of Biology* 31, 155–78.

22. Fitch, W.M. and Margoliash, E. 1967. Construction of Phylogenetic Trees. *Science* 155, 279–84.

23. In 1998, Cambridge researchers Christopher Howe, Adrian Barbrook, and colleagues provided unique insight into the history of more than 50 fifteenth-century versions of Geoffrey Chaucer's *The Canterbury Tales* using the very same analytical procedures used to analyse DNA sequences. Barbrook, A.C., Howe, C.J., Blake, N., and Robinson, P. 1998. The Phylogeny of *The Canterbury Tales*. *Nature* 294, 839.

24. Burkhardt, F. and Smith, S. 1990. *The Correspondence of Charles Darwin 1856–1857*. Vol. 6. Cambridge University Press, p. 456.

3. The Seeds of Symbiosis

1. Honneger, R. 2002. Simon Schwendener (1829–1919) and the Dual Hypothesis of Lichens. *The Bryologist* 103, 307–13.

2. Sapp, J. 1994. *Evolution by Association (A History of Symbiosis)*. Oxford University Press.

3. De Bary, A. 1879. Die Erscheinung der Symbiose. Strasbourg.

4. Mereschkowsky, C. 1905. Über Natur und Ursprung der Chromatophoren im Pflanzenreiche. *Biologisches Centralblatt* 25, 593–604. English translation in Martin, W., Kowallik, K.V. 1999. Annotated English translation of Mereschkowsky's 1905 paper 'Über Natur und Ursprung der Chromatophoren im Pflanzenreiche'. *European Journal of Phycology* 34, 287–95.

5. Page 65 in Mereschkowsky, C. 1920. La Plante Considerée Comme Un Complexe Symbiotique. Societé des Sciences Naturelles de l'Ouest de la France, Nantes, Bulletin. 6, 17–98.

6. Page 112–13 in Schimper A.F.W. 1883. Über die Entwicklung der Chlorophylkörner und Farbkörner. *Botanische Zeitung* 41, 105–14.

7. Mereschkowsky, C. 1905. Über Natur und Ursprung der Chromatophoren im Pflanzenreiche.

8. Consideration of Mereschkowsky's career and the time in which he lived paints a picture of a man conflicted in life and science. He held extreme right-wing sociopolitical views. He was a secret police informant in pre-revolutionary Russia and member of an anti-Semitic organization. A convicted paedophile, Mereschkowsky was something of a vagabond;

ENDNOTES

he spent extended periods of time in the United States (under a false name) and France. He eventually took his own life in ritualistic fashion in a Geneva hotel room, broke and bitter, convinced that his symbiotic theory had made little impact. A fascinating account of the life and times of Mereschkowsky was published by Sapp, Carrapiço, and Zolotonosov. It illustrates the striking disconnect between Mereschkowsky's scientific theories (e.g. symbiogenesis and the notion of 'cooperation' inherent therein) and his eugenics and racism. Sapp, J., Carrapiço, F., and Zolotonosov, M. 2002. Symbiogenesis: The Hidden Face of Constantin Merezhkowsky. *History & Philosophy of the Life Sciences* 24, 413–40.

9. Mereschkowsky, Über Natur und Ursprung der Chromatophoren im Pflanzenreiche.
10. Portier, P. 1918. *Les Symbiotes*. Masson.
11. Sapp, *Evolution by Association (A History of Symbiosis)*, p. 90.
12. Anonymous. 1919. *Nature* 103, 482–3.
13. Lumière, A. 1919. *Le Myth des Symbiotes*. Masson.
14. Wallin, I.E. 1927. *Symbionticism and the Origin of Species*. The Williams & Wilkins Company.
15. It's worth remembering that at this time neither Wallin nor anyone else knew precisely what a gene was. Wallin, I.E. 1925. On the Nature of Mitochondria. IX. Demonstration of the Bacterial Nature of Mitochondria. *The American Journal of Anatomy* 36, 131–49.
16. Unlike Portier, it is clear that when Wallin stated that he had cultured mitochondria he meant it literally. For someone who essentially had his back up against the same experimental wall, Wallin was highly critical of Portier's work. Jan Sapp believes that Wallin, who published in English, was responsible for the spread of much misinformation about Portier and his French publications. Sapp, *Evolution by Association (A History of Symbiosis)*.
17. Mehos, D.C. 1992. Appendix. In Khakhina, L. 1979. *Concepts of Symbiogenesis: A Historical and Critical Study of the Research of Russian Botanists*. Trans. Merkel, S. and Coalson, R. Yale University Press.
18. Gould, S.J. 1989. *A Wonderful Life*. W.W. Norton and Company.
19. 'Evolution by jerks', as John Turner referred to it in a New Scientist article, as opposed to 'evolution by creeps' retorted the Gould camp. Turner, J. 1984. Why We Need Evolution By Jerks. *New Scientist* Feb. 9, 34–5.
20. Morgan, T.H. 1926. Genetics and the Physiology of Development. *American Naturalist* 60, 489–515.
21. Wilson, E.B. 1925. *The Cell in Development and Heredity*. MacMillan, pp. 738–9.
22. Sagan, L. 1967. On the Origin of Mitosing Cells. *Journal of Theoretical Biology* 14, 225–74; Margulis, L. 1998. *Symbiotic Planet (A New Look at Evolution)*. Basic Books.
23. Margulis, L. 1970. *Origin of Eukaryotic Cells*. Yale University Press.
24. Brockman, J. 1995. *The Third Culture*. Touchstone, p. 129. See also Margulis, *Symbiotic Planet (A New Look at Evolution)*.

25. Mann, C. 1991. Lynn Margulis: Science's Unruly Earth Mother. *Science* 252, 378–81.

26. In the late 1980s Margulis received a fellowship that would allow her to delve into the early research of the Russian 'symbiogeneticists'. But she did not read or speak Russian, and to her 'surprise and relief' she discovered that a Russian historian of science named Liya Nikolaevna Khakhina had already written authoritatively on the subject. Margulis dutifully facilitated the translation from Russian into English of Khakhina's 1979 book (see Margulis, L. and McMenamin, M. 1992. Editor's Introduction. In Khakhina, *Concepts of Symbiogenesis*). Margulis and her colleagues did the same in 2010 for a book published in Russian in 1924 by Boris Kozo-Polyansky (1890–1957) (Kozo-Polyansky, B. 1924. *Symbiogenesis: A New Principle of Evolution*. Trans. Fet, V. Harvard University Press). Margulis first learned of Kozo-Polyansky at the 1975 International Botanical Congress in Leningrad—she and her fellow American Peter H. Raven were handed a partial translation of Kozo-Polyansky's book by the famous (and at the time annoyed) Russian botanist Armen Takhtajan (1910–2009): 'Here, read this', he scolded them. 'You Anglophones believe all science must be English or German...Kozo-Polyansky published your ideas long before you were born!'

27. Margulis, *Symbiotic Planet (A New Look at Evolution)*, p. 27.

28. Here she used the name L. Alexander Sagan. Plaut, W. and Alexander Sagan, L. 1958. Incorporation of Thymidine in the Cytoplasm of *Amoeba proteus*. *Journal of Biophysical and Biochemical Cytology* 4, 843–6.

29. Sagan, L. 1965. An Unusual Pattern of Tritiated Thymidine Incorporation in *Euglena. Journal of Protozoology* 12, 105–9.

30. Ris, H. and Plaut, W. 1962. Ultrastructure of DNA-containing Areas in the Chloroplast of *Chlamydomonas. Journal of Cell Biology* 13, 383–91.

31. Nass, M.M. and Nass, S. 1963. Intramitochondrial Fibers with DNA Characteristics. I. Fixation and Electron Staining Reactions. *The Journal of Cell Biology* 19, 593–611; Nass, S. and Nass, M.M. 1963. Intramitochondrial Fibers with DNA Characteristics. II. Enzymatic and Other Hydrolytic Treatments. *The Journal of Cell Biology* 19, 613–29.

32. Published in the April 2011 issue of *Discover* magazine.

33. Margulis, L. and Sagan, D. 2002. *Acquiring Genomes: A Theory of the Origins of Species*. Basic Books.

34. Margulis, L., Maniotis, A., MacAllister, J., Scythes, J., Brorson, O., Hall, J., Krumbein, W.E., and Chapman, M.J. 2009. Spirochete Round Bodies. Syphilis, Lyme Disease and AIDS: Resurgence of 'the Great Imitator'? *Symbiosis* 47, 51–8.

35. Bybee, J. 2012. No Subject Too Sacred. In *Lynn Margulis: The Life and Legacy of a Scientific Rebel*. Sagan, D. ed. Chelsea Green Publishing, p. 159.

4. Molecular Rulers of Life's Kingdoms

1. Silverberg, R. 1968. *Hawksbill Station*. Doubleday Books. The book is an expanded version of a shorter novella of the same name published in 1967.
2. Conway Morris, S. 2000. The Cambrian 'Explosion': Slow-Fuse or Megatonnage? *Proceedings of the National Academy of Sciences USA* 97, 4426–9.
3. Butterfield, N. 2011. Terminal Developments in Ediacaran Embryology. *Science* 334, 1655–6.
4. Ribosomes in both prokaryotes and eukaryotes contain three different types of rRNA: large-subunit rRNA, small-subunit rRNA, and so-called '5S' rRNA. Eukaryotes have a fourth kind, '5.8S', which is technically considered part of the large subunit. These details need not concern us here.
5. Woese, C.R. 2005. Q & A. *Current Biology* 15, R111–12.
6. Woese, Q & A.
7. *Bergey's Manual* has evolved steadily over time. The original *Bergey's Manual of Determinative Bacteriology* became the manual of 'Systematic Bacteriology' in 1980. It presently exists as a massive five-volume set with a structure that reflects known or predicted taxonomic relationships.
8. Sapp, J. 2009. *The New Foundations of Evolution (On the Tree of Life)*. Oxford University Press, p. 156.
9. Sanger, F., Brownless, G.G., and Barrell, B.G. 1965. A Two-dimensional Fractionation Procedure for Radioactive Nucleotides. *Journal of Molecular Biology* 13, 373–98.
10. Woese's earliest datasets were comprised of so-called 5S rRNA sequences, the smallest of the rRNAs. He soon began collecting small-subunit rRNA sequences, which are much longer than 5S rRNA (~1,600 nucleotides in bacteria) and thus contain more information for making evolutionary inferences.
11. Woese, C.R., and Fox, G.E. 1977. Phylogenetic Structure of the Prokaryotic Domain: The Primary Kingdoms. *Proceedings of the National Academy of Sciences USA* 74, 5088–90.
12. Mayr, E. 1998. Two Empires or Three? *Proceedings of the National Academy of Sciences USA* 95, 9720–3.
13. Woese, C.R. 1998. Default Taxonomy: Ernst Mayr's View of the Microbial World. *Proceedings of the National Academy of Sciences USA* 95, 11043–6.

5. Bacteria Become Organelles: An Insider's Take

1. Woese, C.R. 1998. Default Taxonomy: Ernst Mayr's View of the Microbial World. *Proceedings of the National Academy of Sciences USA* 95, 11043–6.
2. Interview with Linda Bonen by the author, 3–4 March 2013. Numerous researchers deserve mention for their contributions to Woese's early research, too many to discuss in detail here. In addition to Bonen and

those mentioned in the previous chapter, these include William Balch, Norman Pace, and Mitchell Sogin.

3. Margulis, L. 1970. *Origin of Eukaryotic Cells.* Yale University Press.

4. Interview with Linda Bonen.

5. Doolittle, W.F. 2004. Q & A. *Current Biology* 14, R176–7.

6. Doolittle, W.F. 1993. Sol's World, the RNA World, Our World. *FASEB Journal* 7, 1–2.

7. Interviews with Ford Doolittle by the author, 24 October 2012 and 27 June 2013.

8. Email interview with Christopher Helleiner, 4 July 2013. Doolittle's official job offer came from Helleiner, Professor Emeritus of Biochemistry and Department Head between 1964–1979. During the interview, Helleiner asked the candidate what he would do if he did not get the position. Doolittle responded: 'become a science fiction writer'. Helleiner believed him.

9. Interviews with Ford Doolittle.

10. Barnett, W.E. and Brown, D.H. 1966. Mitochondrial Transfer Ribonucleic Acids. *Proceedings of the National Academy of Sciences USA* 57, 452–8.

11. Interview with Michael Gray by the author, 24 October 2012; email interviews with Michael Gray by the author, 5 November 2012 and 10 July 2013.

12. Interview with Michael Gray.

13. Interviews with Ford Doolittle.

14. Gøksoyr, J. 1967. Evolution of Eucaryotic Cells. *Nature* 214, 1161.

15. Raven, P.H. 1970. A Multiple Origin for Plastids and Mitochondria. *Science* 169, 641–6.

16. Raff, R.A. and Mahler, H.R. 1972. The Non Symbiotic Origin of Mitochondria. *Science* 177, 575–82; Allsopp, A. 1969. Phylogenetic Relationships of the Procaryotes and the Origin of the Eucaryotic Cell. *New Phytologist* 68, 591–612.

17. Interview with F.J.R. 'Max' Taylor by the author, 27 July 2013.

18. Interview with F.J.R. 'Max' Taylor.

19. Taylor, F.J.R. 1974. Implications and Extensions of the Serial Endosymbiosis Theory of the Origin of Eukaryotes. *Taxon* 23, 229–58. This paper stems from Taylor's presentation at the 1st International Congress of Systematic and Evolutionary Biology, which was held 4–12 August 1973 in Boulder, Colorado. Margulis, then at Boston University, organized a symposium entitled Origin and Evolution of the Eukaryotic Cell.

20. Raven, A Multiple Origin for Plastids and Mitochondria; Taylor, Implications and Extensions.

21. Stanier, R.Y. 1970. Some Aspects of the Biology of Cells and Their Possible Evolutionary Significance. *Symposium of the Society for General Microbiology* 20, 1–38.

22. Stanier, R.Y., Douderoff, M., and Adelberg, E. 1963. *The Microbial World*. 2nd edition. Prentice-Hall, p. 85.

23. Klein, R.M. and Cronquist, A. 1967. A Consideration of the Evolutionary and Taxonomic Significance of Some Biochemical, Micromorphological and Physiological Characters in the Thallophytes. *Quarterly Review of Biology* 42, 105–296.

24. Klein, R.M. 1970. Relationships between Blue-green and Red Algae. *Annals of the New York Academy of Science* 175, 623–32.

25. Margulis, *Origin of Eukaryotic Cells*; Margulis, L. 1981. *Symbiosis in Cell Evolution*. W.H. Freeman and Company.

26. Carbon, P., Ehresmann, C., Ehresmann, B., and Ebel, J.P. 1978. The Sequence of *Escherichia coli* Ribosomal 16 S RNA Determined by New Rapid Gel Methods. *FEBS Letters* 94, 152–6.

27. Interviews with Ford Doolittle.

28. Ris, H. and Plaut, W. 1962. Ultrastructure of DNA-containing Areas in the Chloroplast of *Chlamydomonas*. *Journal of Cell Biology* 13, 383–91.

29. Klein and Cronquist, A Consideration.

30. Bonen, L. and Doolittle, W.F. 1975. On the Prokaryotic Nature of Red Algal Chloroplasts. *Proceedings of the National Academy of Sciences USA* 72, 2310–14.

31. Zablen, L.B., Kissil, M.S., Woese, C.R., and Buetow, D.E. 1975. Phylogenetic Origin of the Chloroplast and Prokaryotic Nature of Its Ribosomal RNA. *Proceedings of the National Academy of Sciences USA* 72, 2418–22.

32. Bonen, L. and Doolittle, W.F. 1976. Partial Sequences of 16S rRNA and the Phylogeny of Blue-green Algae and Chloroplasts. *Nature* 261, 669–73.

33. John, P. and Whatley, F.R. 1975. *Paracoccus denitrificans* and the Evolutionary Origin of the Mitochondrion. *Nature* 254, 495–8.

34. Gould, L.J. 1894. Notes on the Minute Structure of *Pelomyxa palustris* (Greef). *Quarterly Journal of Microscopical Science* s2–36, 295–305.

35. John and Whatley, *Paracoccus denitrificans*.

36. Margulis, L. 1970. *Origin of Eukaryotic Cells*.

37. Interview with Philip John by the author, 10 December, 2012.

38. Interview with Michael Gray.

39. Interview with Michael Gray.

40. Interviews with Ford Doolittle and interview with Michael Gray.

41. Bonen, L., Cunningham, R.S., Gray, M.W., and Doolittle, W.F. 1977. Wheat Embryo Mitochondrial 18S Ribosomal RNA: Evidence for Its Prokaryotic Nature. *Nucleic Acids Research* 4, 663–71.

42. Stanier, Some Aspects, 30–31.

43. Interview with Linda Bonen and interviews with Ford Doolittle.

44. Fredrick, J.F. 1981. Evolutionary Theories: Dogmas or Speculations? *Annals of the New York Academy of Sciences* 361, ix–x. The transcripts of the question-and-answer sessions that took place after each speaker's presentation can be read

at the end of their respective publications. The 'juicy bits' appear to have been omitted.

45. Uzzell, T. and Spolsky, C. 1981. Two Data Sets: Alternative Explanations and Interpretations. *Annals of the New York Academy of Sciences* 361, 481–99.

46. Cronquist, A. 1981. Discussion Paper. *Annals of the New York Academy of Sciences* 361, 500–4.

47. Taylor, Implications and Extensions.

48. Margulis, *Origin of Eukaryotic Cells*.

49. Published claims of the discovery of flagellum-associated DNA—more specifically in the so-called 'basal body' region where the flagellum attaches to the cell—have proven unfounded. The following article provides an excellent overview of this contentious area of genetic and cell biological research: Sapp, J. 1998. Freewheeling Centrioles. *History & Philosophy of the Life Sciences* 20, 255–90.

50. Gray, M.W. and Doolittle, W.F. 1982. Has the Endosymbiont Hypothesis Been Proven? *Microbiological Reviews* 46, 1–42.

51. Schwarz, Z. and Kössel, H. 1980. The Primary Structure of 16S rDNA From *Zea mays* Chloroplast is Homologous to *E. coli* 16S rRNA. *Nature* 283, 739–42.

52. Gray and Doolittle, Has the Endosymbiont Hypothesis Been Proven?; Merchant, S.S. 2009. Lawrence Bogorad: Biographical Memoirs. *National Academy of Sciences (U.S.)*, 92; Gray, J.C. 2000. Chloroplast Cytochromes: Discovery and Characterization. In *Discoveries in Plant Biology*. Kung, S.-D. and Yang, S.-F. eds. World Scientific, pp. 275–90.

53. Yang, D., Oyaizu, Y., Oyaizu, H., Olsen, G.J., and Woese, C.R. 1985. Mitochondrial Origins. *Proceedings of the National Academy of Sciences USA* 82, 4443–7.

54. Gray, M.W. 1998. Rickettsia, Typhus and the Mitochondrial Connection. *Nature* 396, 109–10.

55. Taylor, F.J.R. 1999. Ultrastructure as a Control for Protistan Molecular Phylogeny. *American Naturalist* 154, S125–36. Thomas Kuhn (1922–1996) was an American physicist-philosopher who published an influential book entitled *The Structure of Scientific Revolutions*. In it he introduced the term 'paradigm shift'. It's an overused phrase these days, but it does seem appropriate in the case of endosymbiotic theory.

56. Schwartz, R.M., and Dayhoff, M.O. 1978. Origins of Prokaryotes, Eukaryotes, Mitochondria, and Chloroplasts. *Science* 199, 395–403.

6. The Complex Cell: When, Who, Where, and How?

1. Delbrück, M. 1966. A Physicist Looks at Biology. Reprinted in *Phage and the Origins of Molecular Biology*. Cairns, J., Stent, G.S., and Watson, J.D., eds. Cold Spring Harbor Laboratory Press. Original source: 1949. *The Transactions of the Connecticut Academy of Arts and Sciences* 38, 173–90.

2. Jeon, K.W. 1972. Development of Cellular Dependence on Infective Organisms: Micrurgical Studies in Amoebas. *Science* 176, 1122–3.

3. Jeon, K.W. 2004. Genetic and Physiological Interactions in the Amoeba-Bacteria Symbiosis. *Journal of Eukaryotic Microbiology* 51, 502–8. Molecular phylogenetic studies have shown that Jeon's X-bacteria belong to the bacterial genus *Legionella*, a troublesome lineage that includes *L. pneumophila*, the causative agent of Legionnaires Disease (legionellosis). The natural home of *L. pneumophila* in the environment is within amoebae. *L. pneumophila*-containing amoebae are a common contaminant of swimming pools and domestic water systems; if airborne droplets containing *L. pneumophila* are inhaled, serious (and sometimes fatal) respiratory infection can occur in immunocompromised individuals and the elderly.

4. de Duve, C. 2005. *Singularities: Landmarks on the Pathways of Life*. Cambridge University Press.

5. Holland, H.D. 2006. The Oxygenation of the Atmosphere and Oceans. *Philosophical Transactions of the Royal Society, Series B* 361, 903–15.

6. Margulis, L. and Sagan, D. 1986. *Microcosmos: Four Billion Years of Microbial Evolution*. University of California Press.

7. Lane, N. 2002. *Oxygen: The Molecule that Made the World*. Oxford University Press.

8. Slater, E.C. 1994. Peter Dennis Mitchell. 29 September 1920–10 April 1992. *Biographical Memoirs of Fellows of The Royal Society* 40, 283–305.

9. Mitchell, P. 1961. Coupling of Phosphorylation to Electron and Hydrogen Transfer by a Chemi-osmotic Type of Mechanism. *Nature* 191, 144–8.

10. Lane, N. 2005. *Power, Sex, Suicide: Mitochondria and the Meaning of Life*. Oxford University Press.

11. Schopf, W.J. 1999. *Cradle of Life: The Discovery of Earth's Earliest Fossils*. Princeton University Press.

12. Lepot, K., Benzerara, K., Brown Jr, G.E., and Philippot, P. 2008. Microbially Influenced Formation of 2,724-million-year-old Stromatolites. *Nature* 1, 118–21.

13. Porter, S.M. and Knoll, A.H. 2000. Testate Amoebae in the Neoproterozoic Era: Evidence from Vase-shaped Microfossils in the Chuar Group, Grand Canyon. *Paleobiology* 26, 360–85; Porter, S.M., Meisterfeld, R., and Knoll, A.H. 2003. Vase-shaped Microfossils from the Neoproterozoic Chuar Group, Grand Canyon: A Classification Guided by Modern Testate Amoebae. *Journal of Paleontology* 77, 409–29.

14. Buick, R. 2010. Early Life: Ancient Acritarchs. *Nature* 463, 885–6.

15. In case you are wondering, Russell and Ford are related: 'Russell F. Doolittle and I recently have ascertained that we descend from a remote common ancestral couple—Ebeneezer and Hannah (nee Hall) Doolittle—via eight intermediate nodes on Russell's side and seven, including two more

Ebeneezers, on mine'. Doolittle, W.F. 1997. Fun with Genealogy. *Proceedings of the National Academy of Sciences USA* 97, 4426–9.

16. Doolittle, R.F., Feng, D.F., Tsang, S., Cho, G., and Little, E. 1996. Determining Divergence Times of the Major Kingdoms of Living Organisms with a Protein Clock. *Science* 271, 470–7.

17. Hasegawa, M., Fitch, W.M., Gogarten, J.P., Olendzenski, L., Hilario, E., Simon, C., Holsinger, K.E., Doolittle, R.F., Feng, D.F., Tsang, S., Cho, C., and Little, E. 1996. Dating the Cenancester of Organisms. *Science* 274, 1750–3.

18. Martin, W. 1996. Is There Something Wrong With the Tree of Life? *BioEssays* 18, 523–7.

19. It is also clear that different genes within the same genome can 'tick' at a different rate. For this reason researchers aim to analyse as many genes/proteins as possible and calculate an average degree of divergence among the sequences for a given set of organisms. For an overview of the complexities of molecular clock research, see Bromham, L. and Penny, D. 2003. The Modern Molecular Clock. *Nature Reviews (Genetics)* 4, 216–24.

20. Interview with Tom Cavalier-Smith by the author, 11 December 2012.

21. Sagan, L. 1967. On the Origin of Mitosing Cells. *Journal of Theoretical Biology* 14, 225–74.

22. de Duve, C. 1969. Evolution of the Peroxisome. *Annals of the New York Academy of Sciences* 168, 369–81; Stanier, R.Y. 1970. Some Aspects of the Biology of Cells and Their Possible Evolutionary Significance. *Symposium of the Society for General Microbiology* 20, 1–38.

23. Cavalier-Smith, T. 1983. A 6-Kingdom Classification and Unified Phylogeny. In *Endocytobiology II.* Schwemmler, W. and Schenk, H.E.A. eds. De Gruyter, Berlin, pp. 1027–34.

24. Bovee, E.C. and Jahn, T.L. 1973. Taxonomy and Phylogeny. In *The Biology of Amoeba.* Jeon, K. ed. Academic Press, New York, pp. 38–76.

25. Sogin, M.L., Gunderson, J.H., Elwood, H.J., Alonso, R.A., and Peattie, D.A. 1989. Phylogenetic Meaning of the Kingdom Concept: An Unusual Ribosomal RNA from *Giardia lamblia. Science* 243, 75–7; Vossbrinck, C.R., Maddox, J.V., Friedman, S., Debrunner-Vossbrinck, B.A., and Woese, C.R. 1987. Ribosomal RNA Sequence Suggests Microsporidia Are Extremely Ancient Eukaryotes. *Nature* 326, 411–14.

26. Roger, A.J. 1999. Reconstructing Early Events in Eukaryotic Evolution. *American Naturalist* 154, S146–63; Embley, T.M. and Martin, W. 2006. Eukaryotic Evolution, Changes and Challenges. *Nature* 440, 623–30.

27. Roger, Reconstructing Early Events in Eukaryotic Evolution.

28. Woese, C. 1977. Endosymbionts and Mitochondrial Origins. *Journal of Molecular Evolution* 10, 93–6; Andersson, S.G.E. and Kurland, C.G. 1999. Origins of Mitochondria and Hydrogenosomes. *Current Opinions in Microbiology* 2, 535–41.

29. In cases where a group of organisms does not have a particular feature—a certain subset of proteins involved in the endomembrane, for example—a strong case can invariably be made for recent loss or modification. This parallels the situation for the evolution of MROs from mitochondria. For an overview of eukaryotic cell evolution inferred from genomic data, see Koonin, E.V. 2010. The Origin and Early Evolution of Eukaryotes in the Light of Phylogenomics. *Genome Biology* 11, 209.

30. See for example Yamaguchi, M., Mori, Y., Kozuka, Y., Okada, H., Uematsu, K., Tame, A., Furukawa, H., Maruyama, T., Worman, C.O., and Yokayama, K. 2012. Prokaryote or Eukaryote? A Unique Microorganism from the Deep Sea. *Journal of Electron Microscopy* 61, 423–31. Based on the limited structural information currently in hand it is not clear what this organism is.

31. Note that not all archezoans have been investigated as thoroughly as *Giardia*, *Entamoeba*, and the microsporidian *Vairimorpha*. The amoeba *Pelomyxa*, for example, has not been looked at sufficiently closely to determine whether it has a mitochondrion-related organelle (MRO). However, based on its closest relatives on the tree of eukaryotes it would be shocking if it did not. Researchers are thus confident in concluding that all eukaryotes have mitochondria or MROs.

32. O'Malley, M. 2010. The First Eukaryote Cell: An Unfinished History of Contestation. *Studies in History and Philosophy of Biological and Biomedical Sciences* 41, 212–24.

33. Interviews with William Martin by the author, 2–4 December 2012 and 18–21 August 2013.

34. Müller's initial hydrogenosome research was actually carried out using *Tritrichomonas*, a close relative of *Trichomonas*. Lindmark, D.G. and Müller, M. 1973. Hydrogenosome, a Cytoplasmic Organelle of The Anaerobic Flagellate *Tritrichomonas foetus*, and Its Role in Pyruvate Metabolism. *Journal of Biological Chemistry* 248, 7724–8.

35. Danovaro, R., Dell'Anno, A., Pusceddu, A., Gambi, C., Heiner, I., and Møbjerg, K. 2010. The First Metazoa Living in Permanently Anoxic Conditions. *BMC Biology* 8, 30.

36. Boxma, B., de Graaf, R.M., van der Staay, G.W., van Alen, T.A., Ricard, G., Gabaldón, T., van Hoek, A.H., Moon-van der Staay, S.Y., Koopman, W.J., van Hellemond, J.J., Tielens, A.G., Friedrich, T., Veenhuis, M., Huynen, M.A., and Hackstein, J.H. 2005. An Anaerobic Mitochondrion that Produces Hydrogen. *Nature* 434, 74–9.

37. Stechmann, A., Hamblin, K., Perez-Brocal, V., Gaston, D., Richmond, G.S., van der Giezen, M., Clark, C.G., and Roger, A.J. 2008. Organelles in *Blastocystis* that Blur the Distinction between Mitochondria and Hydrogenosomes. *Current Biology* 18, 580–5.

38. Martin, W. and Müller, M. 1998. The Hydrogen Hypothesis for the First Eukaryote. *Nature* 392, 37–41.

39. The ox-tox and hydrogen hypotheses are prominent examples of the two basic sorts of models for eukaryogenesis to be found in the primary literature. For the sake of simplicity we shall limit ourselves to these two. For detailed discussion and references related to the various models of eukaryotic evolution, readers are encouraged to consult the following articles: Embley and Martin, Eukaryotic Evolution, Changes and Challenges; O'Malley, The First Eukaryote Cell; and de Duve, C. 2007. The Origin of Eukaryotes: A Reappraisal. *Nature Reviews Genetics* 8, 395–403.

40. There are a few examples of 'giant' bacteria in nature. The specific details of how they manage to sustain themselves, bioenergetically speaking, serve as 'exceptions that prove the rule'. We needn't concern ourselves with the details.

41. Lane, N. and Martin, W. 2010. The Energetics of Genome Complexity. *Nature* 467, 929–34.

7. Green Evolution, Green Revolution

1. As mentioned in Chapter 3, the terms 'chloroplast' and 'plastid' are often used interchangeably in reference to the cyanobacterium-derived organelles of plants and algae. Technically speaking, however, it should be noted that while all chloroplasts are plastids, not all plastids are chloroplasts. In plants, for example, the term chloroplast refers specifically to the green-pigmented, actively photosynthesizing version of the organelle, which is in fact only one of several biochemically distinct types of plastid that can exist in the organism. As we shall see in this chapter, when the full breadth of algal diversity is considered, plastids are remarkably diverse in structure and pigmentation.

2. Knoll, A. 2003. *Life On A Young Planet*. Princeton University Press.

3. For those wanting to learn more about the evolution of photosynthesis, I happily recommend the following two books. The first is *Eating the Sun: How Plants Power the Planet* by Oliver Morton (2008, Harper) and the second is Nick Lane's *Life Ascending: The Ten Great Inventions of Evolution* (2009, Norton). Morton's book is science and poetry wrapped in one. Lane's chapter entitled Photosynthesis is a highly readable account of how the photosystems in the earliest prokaryotes might have arisen.

4. Morton, *Eating the Sun*, p. 86.

5. Bonen, L. and Doolittle, W.F. 1975. On the Prokaryotic Nature of Red Algal Chloroplasts. *Proceedings of the National Academy of Sciences USA* 72, 2310–14; Zablen, L.B., Kissil, M.S., Woese, C.R., and Buetow, D.E. 1975. Phylogenetic Origin of the Chloroplast and Prokaryotic Nature of Its Ribosomal RNA. *Proceedings of the National Academy of Sciences USA* 72, 2418–22; Bonen, L. and Doolittle, W.F. 1976. Partial Sequences of 16S rRNA and the Phylogeny of Blue-green Algae and Chloroplasts. *Nature* 261, 669–73.

6. Bonen, L., Cunningham, R.S., Gray, M.W., and Doolittle, W.F. 1977. Wheat Embryo Mitochondrial 18S Ribosomal RNA: Evidence for Its Prokaryotic Nature. *Nucleic Acids Research* 4, 663–71.

7. Sanger, F., Nicklen, S., and Coulson, A.R. 1977. DNA Sequencing with Chain-terminating Inhibitors. *Proceedings of the National Academy of Sciences USA* 74, 5463–7.

8. There were in fact several different DNA sequencing methods floating around at the time. In 1977 Allan Maxam and Walter Gilbert of Harvard University published a 'chemical sequencing' procedure that represented a significant advance over earlier techniques, including Sanger's initial DNA-based methods (Maxam, A.M. and Gilbert, W. 1977. A New Method for Sequencing DNA. *Proceedings of the National Academy of Sciences USA* 74, 560–4). 'Maxam–Gilbert' sequencing was, however, more complicated than Sanger's 1977 'chain-termination' approach and was not amenable for use in a high-throughput format. The rest, as they say, is history. Sanger sequencing was the method of choice for the better part of 30 years.

9. The human genome was sequenced and published twice by different research teams supported by public and private funds. The references for these landmark papers are as follows: International Human Genome Sequencing Consortium. 2001. Initial Sequencing and Analysis of the Human Genome. *Nature* 409, 860–921; Venter, J.C. et al. 2001. The Sequence of the Human Genome. *Science* 291, 1304–51.

10. To scientists educated after about 1980 it is difficult to fathom using a microscope to estimate the size of a genome. But it was remarkably accurate. For example, in 1975 Reinhold Herrmann, Klaus Kowallik, and colleagues at the University of Düsseldorf in Germany showed that the chloroplast DNA of spinach was a circular molecule with a circumference of ~45 micrometres. Knowing the physical dimensions of the DNA double helix and the average mass of a single base pair (A-T or G-C), the authors estimated the mass of a single spinach plastid DNA molecule to be ~1.0 e-8 daltons. This corresponds to a length of ~155,000 base pairs. Years later, when the spinach chloroplast genome was sequenced by Christian Schmitz-Linneweber et al. it was found to be 150,725 base pairs. See Herrmann, R.G., Bohnert, H.-J., Kowallik, K.V., and Schmitt, J.M. 1975. Size, Conformation and Purity of Chloroplast DNA of Some Higher Plants. *Biochimica et Biophysica Acta* 378, 305–17; Hollenberg, C.P., Borst, P., and Van Bruggen, E.F.J. 1970. Mitochondrial DNA V. A 25-μ Closed Circular Duplex DNA Molecule In Wild-Type Yeast Mitochondria: Structure and Genetic Complexity. *Biochimica et Biophysica Acta* 209, 1–15; Petes, T.D., Byers, B., and Fangman, W.I. 1973. Size and Structure of Yeast Chromosomal DNA. *Proceedings of the National Academy of Sciences USA* 70, 3072–6; Manning, J.E., Wolstenholme, D.R., Ryan, R.S., Hunter, J.A., and Richards, O.C. 1971. Circular Chloroplast DNA from *Euglena gracilis*. *Proceedings of the National*

Academy of Sciences USA 68, 1169–73; Schmitz-Linneweber, C., Maier, R.M., Alcaraz, J.P., Cottet, A., Herrmann, R.G., Mache, R. 2001. The Plastid Chromosome of Spinach (*Spinacia oleracea*): Complete Nucleotide Sequence and Gene Organization. *Plant Molecular Biology* 45, 307–15.

11. Anderson, S., Bankier, A.T., Barrell, B.G., de Bruijn, M.H.L., Coulson, A.R., Drouin, J., Eperon, I.C., Nierlich, D.P., Roe, B.A., Sanger, F., Schreier, P.H., Smith, A.J.H., Staden, R., and Young, I.G. 1981. Sequence and Organization of the Human Mitochondrial Genome. *Nature* 290, 457–65.

12. Burger, G., Gray, M.W., Forget, L., and Lang, B.F. 2013. Strikingly Bacteria-like and Gene-rich Mitochondrial Genomes throughout Jakobid Protists. *Genome Biology and Evolution* 5, 418–38.

13. Dagan, T., Roettger, M., Stucken, K., Landan, G., Koch, R., Major, P., Gould, S.B., Goremykin, V.V., Rippka, R., Tandeau de Marsac, N., Gugger, M., Lockhart, P.J., Allen, J.F., Brune, I., Maus, I., Pühler, A., Martin, W.F. 2013. Genomes of Stigonematalean Cyanobacteria (Subsection V) and the Evolution of Oxygenic Photosynthesis from Prokaryotes to Plastids. *Genome Biology and Evolution* 5, 31–44.

14. Ellis, J.R. 1981. Chloroplast Proteins: Synthesis, Transport, and Assembly. *Annual Review of Plant Physiology* 32, 111–37.

15. Weeden, N.F. 1981. Genetic and Biochemical Implications of the Endosymbiotic Origin of the Chloroplast. *Journal of Molecular Evolution* 17, 133–9.

16. Martin, W., Rujan, T., Richly, E., Hansen, A., Cornelsen, S., Lins, T., Leister, D., Stoebe, B., Hasegawa, M., and Penny, D. 2002. Evolutionary Analysis of *Arabidopsis*, Cyanobacterial, and Chloroplast Genomes Reveals Plastid Phylogeny and Thousands of Cyanbacterial Genes in the Nucleus. *Proceedings of the National Academy of Sciences USA* 99, 12246–51.

17. Martin, W. 2010. Evolutionary Origins of Metabolic Compartmentalization in Eukaryotes. *Philosophical Transactions of the Royal Society, Series B* 365, 847–55.

18. Martin, W., Brinkmann, H., Savonna, C., and Cerff, R. 1993. Evidence for a Chimeric Nature of Nuclear Genomes: Eubacterial Origin of Eukaryotic Glyceraldehyde-3-Phosphate Dehydrogenase Genes. *Proceedings of the National Academy of Sciences USA* 90, 8692–6.

19. Huang, C.Y., Ayliffe, M.A., Timmis, J.N. 2003. Direct Measurement of the Transfer Rate of Chloroplast DNA into the Nucleus. *Nature* 422, 72–6; Stegemann, S., Hartmann, S., Ruf, S., and Bock, R. 2003. High-frequency Gene Transfer from the Chloroplast Genome to the Nucleus. *Proceedings of the National Academy of Sciences USA* 100, 8828–33; Martin W. 2003. Gene Transfers from Organelles to the Nucleus: Frequent and in Big Chunks. *Proceedings of the National Academy of Sciences USA* 100, 8612–14.

20. Lister, D.L., Bateman, J.M., Purton, S., and Howe, C.J. 2003. DNA Transfer from Chloroplast to Nucleus is Much Rarer in *Chlamydomonas* than in Tobacco. *Gene* 316, 33–8. It is important to recognize that the results of

Lister et al. apply only to modern-day *Chlamydomonas*; they cannot tell us about rates of EGT in the ancestors of *Chlamydomonas* and other green algae.

21. Mourier, T., Hansen, A.J., Willerslev, E., and Arctander, P. 2001. The Human Genome Project Reveals a Continuous Transfer of Large Mitochondrial Fragments to the Nucleus. *Molecular Biology and Evolution* 18, 1833–7; Tourmen, Y., Baris, O., Dessen, P., Jacques, C., Malthièry, Y., and Teynier, P. 2002. Structure and Chromosomal Distribution of Human Mitochondrial Pseudogenes. *Genomics* 80, 71–7.

22. Timmis, J.N., Ayliffe, M.A., Huang, C.Y., and Martin, W. 2004. Endosymbiotic Gene Transfer: Organelle Genomes Forge Eukaryotic Chromosomes. *Nature Reviews Genetics* 5, 123–35.

23. Aronsson, H. and Jarvis, P. 2009. The Chloroplast Protein Import Apparatus, Its Components, and Their Roles. In *The Chloroplast-Interactions with the Environment*. Aronsson, H. and Sandelius, A.S. eds. Springer-Verlag; Chacinska, A., Koehler, C.M., Milenkovic, D., Lithgow, T., Pfanner, N. 2009. Importing Mitochondrial Proteins: Machineries and Mechanisms. *Cell* 138, 628–44.

24. Aronsson and Jarvis, The Chloroplast Protein Import Apparatus; Chacinska et al., Importing Mitochondrial Proteins.

25. Martin et al., Evolutionary Analysis Of *Arabidopsis*.

26. Huang et al., Direct Measurement; Stegemann et al., High-frequency Gene Transfer; Martin, Gene Transfers From Organelles To The Nucleus.

27. Adams, K.L. and Palmer, J.D. 2003. Evolution of Mitochondrial Gene Content: Gene Loss and Transfer to the Nucleus. *Molecular Phylogenetics and Evolution* 29, 380–95.

28. Recall from Chapter 6 that in the earliest stages of the evolution of the mitochondrion it seems likely that the recipient host genome was not in fact a eukaryotic nuclear genome, but a prokaryotic one. For the sake of simplicity, we have considered EGTs involving mitochondria and chloroplasts together in this chapter. The same general principles of gene transfer and protein import apply to both organelles, although the precise nature of the original host organism may well have been very different.

29. Adams and Palmer, Evolution of Mitochondrial Gene Content.

30. Martin, W. and Schnarrenberger, C. 1997. The Evolution of the Calvin Cycle from Prokaryotic to Eukaryotic Chromosomes: A Case Study of Functional Redundancy in Ancient Pathways through Endosymbiosis. *Current Genetics* 32, 1–18.

31. Respiration-associated DNA damage is one of the reasons why mutation rates are generally higher in organellar genomes than nuclear genomes. All things considered, the nucleus is a 'safer' place for a gene to be.

32. The precise nature of this particular form of organelle-to-nucleus signalling is an area of active research. It appears to involve proteins that are capable of sensing the extent to which electrons are moving down the chain. If a lag is

detected, a biochemical cascade is triggered, a sequential set of interactions between proteins and small molecules, the end result of which is the recruitment of proteins to the nucleus capable of altering the expression of the genes in question.

33. Allen, J.F., Puthiyaveetil, S., Ström, J., and Allen, C.A. 2005. Energy Transduction Anchors Genes in Organelles. *BioEssays* 27, 426–35. See also Allen, J.F. 2003. The Function of Genomes in Bioenergetic Organelles. *Philosophical Transactions of the Royal Society, Series B* 358, 19–38.

34. Delaux, P.-M., Kaur Nanda, A., Mathé, C., Sejalon-Delmas, N., and Dunand, C. 2012. Molecular and Biochemical Aspects of Plant Terrestrialization. *Perspectives in Plant Ecology, Evolution and Systematics* 14, 49–59; Lewis, L.A. and McCourt, R.M. 2004. Green Algae and the Origin of Land Plants. *American Journal of Botany* 91, 1535–56.

35. Butterfield, N.J. 2000. *Bangiomorpha pubescens* n. gen., n. sp.: Implications for the Evolution of Sex, Multicellularity, and the Mesoproterozoic/Neoproterozoic Radiation of Eukaryotes. *Paleobiology* 26, 386–404. Most researchers are inclined to accept that Butterfield's fossils correspond to red algae—Tom Cavalier-Smith is not. He believes that they are more likely to be some sort of filament-forming cyanobacterium. Refer to Chapter 6 for a more detailed discussion of the uncertainties associated with many of the putatively eukaryotic fossils that are more than ~750 million years old.

36. Palmer, J.D. 2003. The Symbiotic Birth and Spread of Plastids: How Many Times and Whodunit? *Journal of Phycology* 39, 4–11; Gould, S., Waller, R.F., and McFadden, G.I. 2008. Plastid Evolution. *Annual Review of Plant Biology* 59, 491–517. Remarkably, tertiary endosymbiosis also takes place. It is conceptually very similar to secondary endosymbiosis in that eukaryotic organisms play the role of both host and endosymbiont.

37. Klein, R.M. and Cronquist, A. 1967. A Consideration of the Evolutionary and Taxonomic Significance of Some Biochemical, Micromorphological and Physiological Characters in the Thallophytes. *Quarterly Review of Biology* 42, 105–296.

38. Raven, P.H. 1970. A Multiple Origin for Plastids and Mitochondria. *Science* 169, 641–6.

39. Schwartz, R.M. and Dayhoff, M.O. 1978. Origins of Prokaryotes, Eukaryotes, Mitochondria, and Chloroplasts. *Science* 199, 395–403.

40. Van Valen, L.M. and Maiorana, V.C. 1980. The Archaebacteria and Eukaryote Origins. *Nature* 287, 248–50.

41. Douglas, S.E. and Turner, S. 1991. Molecular Evidence for the Origin of Plastids from a Cyanobacterium-like Ancestor. *Journal of Molecular Evolution* 33, 267–73; Turner, S., Pryer, K.M., Miao, V.P.W., and Palmer, J.D. 1999. Investigating Deep Phylogenetic Relationships among Cyanobacteria and Plastids by Small Subunit rRNA Sequence Analysis. *Journal of Eukaryotic Microbiology* 4, 327–38; Martin, W., Stoebe, B., Goremykin, V., Hansmann,

S., Hasegawa, M., and Kowallik, K. 1998. Gene Transfer to the Nucleus and the Evolution of Chloroplasts. *Nature* 393, 162–5.

42. McFadden, G.I. and van Dooren, G.G. 2004. Evolution: Red Algal Genome Affirms a Common Origin of All Plastids. *Current Biology* 14, R514–16; Price, D.C. et al. 2012. *Cyanophora paradoxa* Genome Elucidates Origin of Photosynthesis in Algae and Plants. *Science* 335, 843–7.

43. One point of caution is the issue of what biologists call convergent evolution. In this case I am referring to the possibility that the chloroplast genomes in two or more of the three primary chloroplast-bearing groups of algae converged upon roughly similar sizes, structures, and gene sets from *different* cyanobacterial endosymbionts. John Stiller, an American biologist at East Carolina University, is one of the strongest advocates of the possibility of convergent evolution in organellar genome content. On balance, most researchers consider the possibility remote, particularly given the distinct similarities in the chloroplast protein import machineries found in red, green, and glaucophyte algae. Curious readers are referred to Palmer, The Symbiotic Birth and Spread of Plastids, and the following articles by Stiller, Howe, Larkum, and colleagues for discussion of these and related issues: Stiller, J.W., Reel, D.C., and Johnson, J.C. 2003. A Single Origin of Plastids Revisited: Convergent Evolution in Organellar Genome Content. *Journal of Phycology* 39, 95–105; Howe, C.J., Barbrook, A.C., Nisbet, R.E.R., Lockhart, P.J., and Larkum, A.W.D. 2008. The Origin of Plastids. *Philosophical Transactions of the Royal Society, Series B* 363, 2675–85.

44. Parfrey, L.W., Lahr, D.J.G., Knoll, A.H., and Katz, L.A. 2011. Estimating the Timing of Early Eukaryotic Diversification with Multigene Molecular Clocks. *Proceedings of the National Academy of Sciences USA* 108, 13624–9; Yoon, H.S., Hackett, J.D., Ciniglia, C., Pinto, G., and Bhattacharya, D. 2004. A Molecular Timeline for the Origin of Photosynthetic Eukaryotes. *Molecular Biology and Evolution* 21, 809–18.

45. Weber, A.P.M., Linka, M., and Bhattacharya, D. 2006. Single, Ancient Origin of a Plastid Metabolite Translocator Family in Plantae from an Endomembrane-derived Ancestor. *Eukaryotic Cell* 5, 609–12.

46. Weeden, Genetic and Biochemical Implications.

47. Gibbs, S.P. 2006. Looking At Life: From Binoculars to the Electron Microscope. *Annual Reviews of Plant Biology* 57, 1–17.

48. Gibbs's first attempt at publishing the idea of a green algal origin of photosynthesis in *Euglena* ended in failure: 'I submitted a short paper on my theory to *Nature*, which promptly rejected it. Fortunately, the *Canadian Journal of Botany* was kinder to me and published it' (Gibbs, Looking At Life, pp. 13–14). The resulting paper is as follows: Gibbs, S.P. 1978. The Chloroplasts of *Euglena* May Have Evolved from Symbiotic Green Algae. *Canadian Journal of Botany* 56, 2883–9.

49. Taylor, F.J.R. 1974. Implications and Extensions of the Serial Endosymbiosis Theory of the Origin of Eukaryotes. *Taxon* 23, 229–58.

50. An overview of early and current literature on nucleomorphs and their genomes can be found in the following review article: Moore, C. and Archibald, J.M. 2009. Nucleomorph Genomes. *Annual Review of Genetics* 43, 251–64.

51. Douglas, S.E., Zauner, S., Fraunholz, M., Beaton, M., Penny, S., Deng, L., Wu, X., Reith, M., Cavalier-Smith, T., and Maier, U.-G. 2001. The Highly Reduced Genome of an Enslaved Algal Nucleus. *Nature* 410, 1091–6.

52. Gilson, P.R., Su, V., Slamovits, C.H., Reith, M.E., Keeling, P.J., and McFadden, G.I. 2006. Complete Nucleotide Sequence of the Chlorarachniophyte Nucleomorph: Nature's Smallest Nucleus. *Proceedings of the National Academy of Sciences USA* 103, 9566–71.

53. Curtis, B.A. et al. 2012. Algal Nuclear Genomes Reveal Evolutionary Mosaicism and Fate of Nucleomorphs. *Nature* 492, 59–65.

54. Palmer, The Symbiotic Birth and Spread of Plastids; Archibald, J.M. 2009. The Puzzle of Plastid Evolution. *Current Biology* 19, R81–8; Keeling, P.J. 2013. The Number, Speed, and Impact of Plastid Endosymbioses in Eukaryotic Evolution. *Annual Review of Plant Biology* 64, 583–607.

55. Hackett, J.D., Anderson, D.M., Erdner, D.L., and Bhattacharya, D. 2004. Dinoflagellates: A Remarkable Evolutionary Experiment. *American Journal of Botany* 91, 1523–34.

8. Back to the Future

1. Gould, S.J. 1989. *A Wonderful Life*. W.W. Norton and Company, p. 48.

2. Melkonian, M. and Mollenhauer, D. 2005. Robert Lauterborn (1869–1952) and His *Paulinella chromatophora*. *Protist* 156, 253–62.

3. Melkonian, M. and Surek, B. 2009. Famous Algal Isolates from the Spessart Forest (Germany): The Legacy of Dieter Mollenhauer. *Algalogical Studies* 129, 1–23.

4. Bhattacharya, D., Helmchen, T., and Melkonian, M. 1995. Molecular Evolutionary Analyses of Nuclear-encoded Small-subunit Ribosomal RNA Identify an Independent Rhizopod Lineage Containing the Euglyphina and the Chlorarachniophyta. *Journal of Eukaryotic Microbiology* 42, 45–69.

5. Marin, B., Nowack, E.C.M, and Melkonian, M. 2005. A Plastid in the Making: Evidence for a Second Primary Endosymbiosis. *Protist* 156, 425–32.

6. Theissen, U. and Martin, W. 2006. The Difference between Organelles and Endosymbionts. *Current Biology* 16, R1016–17.

7. Cavalier-Smith, T. and Lee, J.J. 1985. Protozoa As Hosts for Endosymbioses and the Conversion of Symbionts into Organelles. *Journal of Protozoology* 32, 376–9.

8. Nowack, E.C., Melkonian, M., and Glöckner, G. (2008). Chromatophore Genome Sequence of *Paulinella* Sheds Light on Acquisition of Photosynthesis by Eukaryotes. *Current Biology* 18, 410–18.

9. Nakayama, T. and Ishida, K.-I. 2009. Another Acquisition of a Primary Photosynthetic Organelle is Underway in *Paulinella chromatophora*. *Current Biology* 19, R284–5.

10. Nowack, E.C., Vogel, H., Groth, M., Grossman, A.R., Melkonian, M., and Glöckner, G. 2011. Endosymbiotic Gene Transfer and Transcriptional Regulation of Transferred Genes in *Paulinella chromatophora*. *Molecular Biology and Evolution* 28, 407–22.

11. Nowack, E.C. and Grossman, A.R. 2012. Trafficking of Protein into the Recently Established Photosynthetic Organelles of *Paulinella chromatophora*. *Proceedings of the National Academy of Sciences USA* 109, 5340–5.

12. Nowack et al., Chromatophore Genome Sequence Of *Paulinella*.

13. Mackiewicz, P., Bodył, A., and Gagat, P. 2012. Possible Import Routes of Proteins into the Cyanobacterial Endosymbionts/Plastids of *Paulinella Chromatophora*. *Theory in Biosciences* 131, 1–18.

14. The details behind this assertion are not necessary for our present discussion. Suffice it to say that both green and red algal endosymbionts have been involved in secondary endosymbioses with different non-photosynthetic host eukaryotes. Another layer of complexity is the phenomenon of tertiary endosymbiosis, whereby a secondary chloroplast-bearing alga (e.g. a diatom) is ingested by another eukaryote. It is possible that one or more tertiary events have contributed to the spread of chloroplasts across the tree of eukaryotes. What is known with certainty is that over short evolutionary timescales and on numerous occasions, the dinoflagellate algae have engaged in tertiary endosymbiosis. Readers wanting to learn more about this 'remarkable evolutionary experiment' are referred to the following article by Bhattacharya and colleagues: Hackett, J.D., Anderson, D.M., Erdner, D.L., and Bhattacharya, D. 2004. Dinoflagellates: A Remarkable Evolutionary Experiment. *American Journal of Botany* 91, 1523–34.

15. Mackiewicz et al., Possible Import Routes of Proteins.

16. Johnson, P.W., Hargraves, P.E., and Sieburth, J.M. 1988. Ultrastructure and Ecology of *Calycomonas ovalis* Wulff, 1919, (Chrysophyceae) and Its Redescription as a Testate Rhizopod, *Paulinella ovalis* N. Comb. (Filosea: Euglyphina). *Journal of Protozoology* 35, 618–26.

17. Prechtl, J., Kneip, C., Lockhart, P., Wenderoth, K., and Maier U.G. 2004. Intracellular Spheroid Bodies of *Rhopalodia gibba* Have Nitrogen-fixing Apparatus of Cyanobacterial Origin. *Molecular Biology and Evolution* 21, 1477–81.

18. Nakayama, T., Ikegama, Y., Nakayama, T., Ishida, K.-I., Inagaki, Y., and Inouye, I. 2011. Spheroid Bodies in Rhopalodiacean Diatoms Were Derived

from a Single Endosymbiotic Cyanobacterium. *Journal of Plant Research* 124, 93–7.

19. Kneip, C., Voss, C., Lockhart, P., and Maier, U.G. 2008. The Cyanobacterial Endosymbiont of the Unicellular Algae *Rhopalodia gibba* Shows Reductive Genome Evolution. *BMC Evolutionary Biology* 8, 30.

20. Wägele, H., Deusch, O., Händeler, K., Martin, R., Schmitt, V., Christa, G., Pinzger, B., Gould, S.B., Dagan, T., Klussmann-Kolb, A., and Martin, W. 2011. Transcriptomic Evidence that Longevity of Acquired Plastids in the Photosynthetic Slugs *Elysia timida* and *Plakobranchus ocellatus* Does Not Entail Lateral Transfer of Algal Nuclear Genes. *Molecular Biology and Evolution* 28, 699–706; Bhattacharya, D., Pelletreau, K.N., Price, D.C., Sarver, K.E., Rumpho, M.E. 2013. Genome Analysis of *Elysia chlorotica* Egg DNA Provides No Evidence for Horizontal Gene Transfer into the Germ Line of This Kleptoplastic Mollusc. *Molecular Biology and Evolution* 30, 1843–52.

21. Also called matryoshka dolls, Russian nesting dolls are colourful children's toys (and popular souvenirs) consisting of a series of successively smaller dolls stacked one inside the other.

22. Larkum, A.W.D., Lockhart, P.J., and Howe, C.J. 2007. Shopping for Plastids. *Trends in Plant Science* 12, 189–95.

23. Baumann, P. 2005. Biology of Bacteriocyte-associated Endosymbionts of Plant Sap-sucking Insects. *Annual Review of Microbiology* 59, 155–89.

24. McCutcheon, J.P. and Moran, N.A. 2011. Extreme Genome Reduction in Symbiotic Bacteria. *Nature Reviews Microbiology* 10, 13–26.

25. Bennett, G.M. and Moran, N.A. 2013. Small, Smaller, Smallest: The Origins and Evolution of Ancient Dual Symbioses in a Phloem-feeding Insect. *Genome Biology and Evolution* 5, 1675–88.

26. Bennett and Moran, Small, Smaller, Smallest.

27. Husnik, F., Nikoh, N., Koga, R., Ross, L., Duncan, R.P., Fujie, M., Tanaka, M., Satoh, N., Bachtrog, D., Wilson, A.C., von Dohlen, C.D., Fukatsu, T., McCutcheon, J.P. 2013. Horizontal Gene Transfer from Diverse Bacteria to an Insect Genome Enables a Tripartite Nested Mealybug Symbiosis. *Cell* 153, 1567–78.

9. Epilogue

1. Berg, P. and Mertz, J.E. 2010. Personal Reflections on the Origins and Emergence of Recombinant DNA Technology. *Genetics* 184, 9–17; Cohen, S. N. 2013. DNA Cloning: A Personal View after 40 Years. *Proceedings of the National Academy of Sciences USA* 110, 15521–9.

2. Berg, P., Baltimore, D., Boyer, H.W., Cohen, S.N., Davis, R.W., Hogness, D.S., Nathans, D., Roblin, R., Watson, J.D., Weissman, S., and Zinder, N.D. 1974. Potential Biohazards of Recombinant DNA Molecules. *Science* 185, 303.

3. Welch, R.A. et al. 2002. Extensive Mosaic Structure Revealed by the Complete Genome Sequence of Uropathogenic *Escherichia coli*. *Proceedings of the National Academy of Sciences USA* 99, 17020–4.

4. Hazkani-Covo, E., Zeller, R.M., and Martin, W. 2010. Molecular Poltergeists: Mitochondrial DNA Copies (*numts*) in Sequenced Nuclear Genomes. *PLoS Genetics* 6, e1000834.

INDEX